Corn Insect Pests: From Biology to Control Technology

Corn Insect Pests: From Biology to Control Technology

Guest Editor

Tiantao Zhang

MDPI

Basel • Beijing • Wuhan • Barcelona • Belgrade • Novi Sad • Cluj • Manchester

Guest Editor
Tiantao Zhang
Institute of Plant Protection
Chinese Academy of
Agricultural Sciences
Beijing
China

Editorial Office
MDPI AG
Grosspeteranlage 5
4052 Basel, Switzerland

This is a reprint of the Special Issue, published open access by the journal *Insects* (ISSN 2075-4450), freely accessible at: https://www.mdpi.com/journal/insects/special_issues/76SBEHZQD5.

For citation purposes, cite each article independently as indicated on the article page online and as indicated below:

Lastname, A.A.; Lastname, B.B. Article Title. *Journal Name* **Year**, *Volume Number*, Page Range.

ISBN 978-3-7258-5607-7 (Hbk)
ISBN 978-3-7258-5608-4 (PDF)
https://doi.org/10.3390/books978-3-7258-5608-4

Contents

insects

MDPI

Article

Bt Trait Efficacy Against Corn Earworm, *Helicoverpa zea*, (Lepidoptera: Noctuidae) for Preserving Grain Yield and Reducing Mycotoxin Contamination of Field Corn

William Yancey Barton [1], George David Buntin [2,*] and Micheal D. Toews [3]

[1] Georgia Department of Agriculture, Atlanta, GA 30334, USA; yancey.barton@agr.georgia.gov
[2] Department of Entomology, University of Georgia-Griffin Campus, Griffin, GA 30223, USA
[3] Department of Entomology, University of Georgia-Tifton Campus, Tifton, GA 31793, USA; mtoews@uga.edu
* Correspondence: gbuntin@uga.edu

Simple Summary: The corn earworm causes persistent ear damage to corn grown in the southeastern United States. Increased levels of ear damage have been associated with mycotoxin contamination such as aflatoxin and fumonisin. Corn hybrids expressing the Bt traits Vip3Aa20 provided substantial corn earworm control and prevented kernel damage. Older Bt corn without this trait did not control corn earworm in the ear and prevent ear damage. Bt corn that prevented kernel damage had a variable effect on grain yield but may prevent yield loss. Bt corn that prevented ear damage did not suffer grain contamination from aflatoxin but did show reduced grain contamination by fumonisin.

Abstract: The corn earworm, *Helicoverpa zea* (Boddie), causes persistent ear damage to corn grown in the southeastern United States region. Increased levels of ear damage have been associated with mycotoxin contamination in addition to yield loss. Corn hybrids expressing proteins from the *Bacillus thuringiensis* (Bt) may provide corn earworm control. A selection of hybrids expressing various Bt traits were evaluated in field experiments across Georgia over two years to assess their efficacy for corn earworm control, grain yield and quality protection, and grain mycotoxin mitigation. Ear damage was significantly reduced only by Bt hybrids expressing the Vip3Aa20 protein. The remaining Bt hybrids expressing Cry proteins provided only marginal control. Ear damage had a variable effect on grain yield and was not correlated with grain aflatoxin contamination. In contrast, grain fumonisin contamination was positively associated with earworm damage. These results indicate Bt hybrids that effectively reduce corn earworm damage may also assist in reducing fumonisin contamination and possibly yield loss.

Keywords: *Zea mays*; transgenic crop; maize; *Bacillus thuringiensis*; fumonisin; aflatoxin; Vip3Aa20

1. Introduction

Corn, *Zea mays* L., genetically modified with proteins from the *Bacillus thuringiensis* (Bt) bacterium, was commercially released in North America as a method of controlling stalk-boring pests including the European corn borer, *Ostrinia nubilalis* (Hübner) (Lepidoptera: Crambidae) [1,2]. This method ultimately provided mixed results when tested against corn earworm, *Helicoverpa zea* (Boddie), a lepidopteran pest species with a wide agronomic host range including field corn. Corn earworm larvae feed on the corn silks, ears and kernels for six instars prior to pupating in the soil, resulting in significant damage to the upper ear region. However, the impact of larval feeding alone is generally not a significant source of yield loss in field corn production [3–5] since most corn earworm activity occurs at the ear tip region where unpollinated kernels are located [6]. However, significant ear damage with some level of grain yield reduction has been reported in late plantings when larval activity is greater [7–10].

Bt traits produce plant proteins and are widely incorporated into transgenic corn hybrids throughout the United States, with 82% of all domestically planted corn in 2020 having expressed single or pyramided Bt toxins [11]. Early studies in the southeastern United States region showed that Cry1Ab in events BT11 and MON810 provided partial control against corn earworm infestations, but significant control and improved yield was mostly observed for whorl infestations by fall armyworm, *Spodoptera frugiperda* (J. E. Smith) (Lepidoptera: Noctuidae) [7,8,12–14]. Additional Bt toxins became commercially available as pyramids with varying levels of efficacy for corn earworm control. The MON89034 event expresses the pyramided proteins Cry1A.105 + Cry2Ab2 and when released provided better control of corn earworm than Cry1Ab [3,8,14–16]. However, resistance to these toxins developed across local lepidopteran pest populations. For corn earworm, resistance to Cry1Ab is widespread and resistance to Cry2Ab was reported at local levels across several southern states [17–20]. The MIR162 event in corn expressing the Vip3Aa20 insecticidal protein currently is the most effective transgenic event for reducing corn earworm numbers in corn [9,18,21–23]. Indeed, currently corn expressing the vip3Aa20 protein almost completely eliminates ear infestation by corn earworm [9,20,24]. Nevertheless, the presence of resistant alleles in corn earworm to Vip3Aa20 protein has been confirmed in localized populations from Texas and the Mid-South. There is evidence that resistance to Vip3Aa20 may be increasing in field-collected corn earworms [24,25].

Lepidopteran damage to corn ears may result in corn becoming vulnerable to toxic secondary metabolites known as mycotoxins [26–29]. The fungus *Aspergillus flavus* (Link) (Deutermycetes: Moniliales) will express the mycotoxin aflatoxin B_1 [29–31]. Corn grain in semi-tropical areas also may be infected by *Fusarium* spp. with *Fusarium verticillioides* (Sacc.) Nirenberg (Sordariomycetes: Hypocreales) being the primary species expressing the mycotoxin fumonisin [28,32]. Grain that is highly contaminated with aflatoxin or fumonisin may be toxic when consumed by humans and animals. A major mycotoxicosis outbreak that occurred across rural Kenya in 2004 was the result of post-harvested corn storage under damp conditions that promoted severe grain aflatoxin contamination, resulting in numerous cases of sickness and death [33]. In addition to excessive moisture during storage, heightened mycotoxin contamination can occur from other factors in the field including high temperatures, drought conditions, and plant genetics [29,34–37]. The U.S. Food and Drug Administration (FDA) limits in grain intended for human consumption are no more than 20 ppb for aflatoxin and 2–4 ppm for fumonisin. Grain intended for animal feed can have higher limits depending on the animal age, weight, and production stages [38,39]; for example, the FDA aflatoxin limit for breeding swine is 200 ppb. Early reports found that increased aflatoxin contamination levels were linked with significant ear injury caused by lepidopteran pests, including corn earworm [26,40]. Increased larval feeding could promote mycotoxin contamination through induced plant stress [30,31,34,37] or the larvae may simply harbor fungal spores within their gut that are transferred into host plants upon feeding [41]. Furthermore, planting date plays a significant role in fungal susceptibility for field corn production in temperate and tropical regions as later planted corn endures more stress [32,34–36].

Both aflatoxin and fumonisin contamination have been associated with ear feeding by larvae of the corn earworm, fall armyworm and several other lepidopterans [31,32,37]. The overall effect of transgenic Bt hybrids in reducing aflatoxin contamination is inconsistent. Multiple studies found significant associations between lepidopteran pest ear damage and aflatoxin contamination levels in transgenic corn [31,34,35,37,42,43]. However, other studies found greater inconsistent results or no clear association despite making similar comparisons [4,7,14,44–46]. Nevertheless, a study using aflatoxin-related insurance claims in the U.S. found that aflatoxin risk was lower in counties where more Bt corn was planted [47]. Although adoption of Bt hybrids may contribute to aflatoxin mitigation [37,47], it remains inconclusive how effective Bt technology is as a tool for aflatoxin reduction in field corn [27].

Conversely, fusarium ear rot and fumonisin contamination have a more definite association with increased ear injury from lepidopteran pest feeding. In Iowa, Munkvold et al. [48] found increased European corn borer injury induced *Fusarium* ear rot, and Bowers et al. [49,50] found positive associations in fumonisin accumulations with increased ear injury from European corn borer, corn earworm and western bean cutworm, *Striacosta albicosta* (Smith) (Lepidoptera: Noctuidae). Western bean cutworm is associated with deoxynivalenol and Gibberella ear rot from accumulated *Fusarium graminearium* infection of corn in the midwestern United States and Ontario, Canada [51,52]. Parson and Munkvold [32] also found a correlation between lepidopteran kernel damage and fumonisin B_1 contamination in corn. A meta-analysis of studies on genetically engineered maize found that hybrids expressing Bt toxins exhibited lower concentrations of mycotoxins by 28.8% and fumonisins by 30.6% [53]. Incorporating Bt technology could potentially reduce grain fumonisin contamination [27,36,52,53], especially use of newer, more effective pyramided hybrids that express the Vip3Aa20 protein [48,52].

The objective was to evaluate how commercially available Bt hybrids expressing various Bt proteins prevented corn earworm ear damage and mycotoxin contamination. A selection of commercially available non-Bt hybrids as well as hybrids expressing various pyramided Bt traits was evaluated. Larval infestations and ear damage were measured and correlated with grain yield loss or grain mycotoxin contamination levels to determine whether Bt technologies preserved grain yield and quality in the field. The authors hypothesized that only those Bt hybrids expressing the Vip3Aa20 protein will effectively reduce corn earworm infestations. Reductions in aflatoxin and fumonisin contamination are expected only where there is significant reduction in corn earworm ear and kernel damage.

2. Materials and Methods

2.1. Field Experiments

Field experiments were conducted at two locations per year in central and southern Georgia in 2019 and 2020. Locations were the University of Georgia Bledsoe Research farm near Griffin (N 33.175964 W −84.409210), the Southwest Georgia Research and Education Center near Plains (N 32.046602 W −84.370610), and the University of Georgia Lang-Rigdon farm near Tifton (N 31.516910 W −83.548479). Soil was an Appling sandy loam in Griffin, a Greenville sandy loam in Plains, and Tifton sandy loam in Tifton. Weed control and fertility practices followed the Georgia Extension Service recommendations for each location. Conventional tillage was used at all locations with chisel plowing followed by disk harrowing. Before disking, 500 kg/ha of a 5–10–15 (N-P-K) granular fertilizer was applied and an additional 112 kg of nitrogen as ammonium nitrate was applied beside the rows and incorporated about 20 days after planting. For weed control, atrazine (Aatrex 4L, Syngenta Crop protection, Greensboro, NC, USA) and pendimethalin (Prowl 3.3EC, BASF, Research Triangle Park, NC, USA) were applied at planting at all locations except in Griffin, where atrazine with acetochlor (Warrant, Bayer CropScience LP, St. Louis, MO, USA) was applied. Plots were treated with a broadcast application of glyphosate (Roundup WeatherMax, Bayer CropScience, St. Louis, MO, USA) about 20 to 25 days after planting for post-emergence weed control. All corn seed was received from the seed companies and pretreated with two or three fungicides and either clothianidin at 0.5 mg per kernel (Poncho 250 or 500, Bayer CropScience, St. Louis, MO, USA) or thiamethoxam at 0.5 mg per kernel (Syngenta Crop Protection, Greensboro, NC, USA). No other insecticides were applied. Natural rainfall was supplemented weekly by irrigation of 6 cm of water as needed.

A selection of available hybrids with various pyramided Bt proteins for above-ground pests were evaluated and compared with non-Bt hybrids of similar relative maturity and agronomic traits (Table 1). Hybrids were provided by DeKalb Seeds (Bayer CropScience, St. Louis, MO, USA) and Pioneer Hi-bred International Inc. (Corteva AgriScience, Johnston, IA, USA). Planting dates in 2019 were 24 April in Plains and 25 April in Tifton, and in 2020 were 3 April in Tifton and 16 May in Griffin. Experimental design for all plantings was a randomized complete block design with four replicates. Corn seed was planted at a rate of

79,040 seeds per ha in 91 cm wide rows at the Plains and Tifton locations and 76 cm wide rows at the Griffin location using a two-row Monosem® pneumatic planter (Largeasse, France). Plots were eight rows wide and 12.2 m long, except in the 2020 Tifton plantings, which were eight rows wide and 10.7 m long.

Table 1. Characteristics of all Bt and non-Bt field corn hybrids used in the 2019 and 2020 plantings.

Brand and Hybrid	Product Name	Bt Toxins [a]	Year
DeKalb DKC 6694	Non-Bt	None	2019, 2020
DeKalb DKC 6697	Genuity VT Double PRO	Cry1A.105, Cry2Ab2	2019, 2020
DeKalb DKC 6629	Genuity Trecepta	Cry1A.105, Cry2Ab2, Vip3Aa20	2019, 2020
DeKalb DKC 6205	Non-Bt	None	2019, 2020
DeKalb DKC 6208	SmartStax	Cry1A.105, Cry2Ab2, Cry1Fa2	2019, 2020
DeKalb DKC 6824	Non-Bt	None	2020
DeKalb DKC 6826	Genuity VT Double PRO	Cry1A.105, Cry2Ab2	2020
DeKalb DKC 6799	Genuity Trecepta	Cry1A.105, Cry2Ab2, Vip3Aa20	2020
Pioneer 1637R	None	None	2019, 2020
Pioneer 1637YHR	Optimum Intrasect	Cry1Ab, Cry1Fa2	2019, 2020
Pioneer 1637VYHR	Optimum Leptra	Cry1Ab, Cry1Fa2, Vip3Aa20	2019
Pioneer 1870R	None	None	2020
Pioneer 1870YHR	Optimum Intrasect	Cry1Ab, Cry1Fa2	2020
Pioneer 2088R	None	None	2019, 2020
Pioneer 2089VYHR	Optimum Leptra	Cry1Ab, Cry1Fa2, Vip3Aa20	2019, 2020

[a] All Bt hybrids also expressed glyphosate herbicide tolerance.

2.2. Data Collection

Plant stand counts were made on the center two rows per plot while inspecting for whorl damage on the corn plants caused by lepidopteran defoliators, primarily fall armyworm larvae, at the five- to eight-leaf vegetative growth stages. The percentage of damaged plants and severity of damage was rated using the Davis et al. [54] 0–9 scale, where 0 represents no damage and 9 represents nearly total destruction of the whorl. Selected infested plants in border rows were inspected to determine the identification of whorl-infesting larvae.

Evaluation for corn earworm infestations occurred when crops reached the milk stage (R3) [55]. Fifteen random ears from each plot were opened from their husks and examined for the presence of corn earworm larvae. The total number of larvae was counted and categorized as either small (first and second instars), medium (third and fourth instars), or large (fifth and sixth instars) in size. Small larvae were identified as no larger than 7 mm, medium larvae were 8 to 24 mm long, and large larvae were greater than 24 mm in length. Exit holes with underlying ear damage also were counted, indicating that a larva had completed development and exited the ear to pupate in the soil. Plots were evaluated for total corn earworm damage at the late dent stage (R5) to early physiological maturity stage (R6) [55]. Another 15 random ears were inspected in each plot and the total area damage from larval feeding was measured in cm^2. Damage measurements for every ear were divided between the tip portion with unpollinated kernels and the rest of the ear containing viable kernels.

At full maturity, grain was harvested from the center two rows not used for earworm sampling of each plot using a two-row self-propelled Wintersteiger Delta combine (Wintersteiger Inc., Salt Lake City, UT, USA) with an automated weighing system that measured plot grain weight, percentage moisture content, and test weight. Grain yields were adjusted to a standard 15.5% moisture content and extrapolated to a kg/ha basis. A 1-kg sample grain was collected from each plot and processed by Waters Agricultural Laboratories, Inc. (Camilla, GA, USA) to measure grain mycotoxin contamination levels. The NEOGEN Veratox® method, a competitive direct ELISA, provided quantitative analyses of total aflatoxin (B_1, B_2, G_1, G_2) (ppb) and total fumonisin (B_1, B_2, B_3) (ppm) contamination levels of corn grain [56,57]. Both methods are approved by the AOAC Research Institute [58].

Samples that read higher levels than the standard provided in each kit were diluted then reanalyzed with the additional dilution factor applied to the quantitative value.

2.3. Statistical Analyses

The statistical software SAS version 9.4 and JMP Pro version 15.0.0 were used for data analysis [59,60]. Results were generalized across locations but were analyzed separately by year because some hybrids exhibited wide variability between 2019 and 2020. Results were statistically analyzed through an analysis of variance (ANOVA) mixed model with PROC MIXED and appropriate statistical procedures for a randomized completed block design including hybrids coded as a fixed effect and replicates coded as a random effect with the Kenward–Roger degrees of freedom approximation option. Before analysis, percentage values were transformed with an angular transformation before analysis and larva counts and damage and mycotoxin values were transformed with a log10(x + 1) transformation to normalize variances. Non-transformed values are presented in the results and figures. Single degree-of-freedom contrasts were used to compare the responses of hybrids expressing Bt proteins with non-Bt hybrids, Bt hybrids expressing only Cry proteins with non-Bt hybrids, Bt hybrids expressing Vip3Aa20 with non-Bt hybrids, and Bt hybrids expressing only Cry proteins with hybrids expressing Cry + Vip3Aa20 proteins. When hybrid treatment F tests were significant, means were separated using pairwise t-test groupings in PROC PLM when a significant difference was indicated among F-values ($\alpha = 0.05$). Linear regression of a bivariate fit of two continuous data type variables using PROC REG [59] was used to test associations of corn earworm kernel damage with grain yield and contamination levels of total grain aflatoxin and fumonisin.

3. Results

3.1. Infestation Rates and Ear Damage

Little whorl defoliation was observed during the vegetative growth stages in all experiments and the small number of non-Bt plants with whorl damage was caused by fall armyworm infestation. Overall corn earworm infestations and damage ratings were lower than normal for field corn in southern Georgia for both 2019 and 2020. Infestation levels were significantly higher in those experiments planted from late April into May than those planted in early April. Corn earworm infestations were significantly different among hybrids in 2019 ($F = 96.98$, df = 9, 54; $p < 0.0001$) and 2020 ($F = 25.85$, df = 11, 66; $p < 0.0001$) (Tables 2 and 3). Overall, Bt hybrids reduced corn earworm infestations during the R3 growth stage by an average of 65.2% in 2019 (non-Bt: 73.12 ± 3.53% infested plants; Bt: 25.42 ± 4.53% infested plants) and 53.6% in 2020 (non-Bt: 53.33 ± 5.08% infested plants; Bt: 24.76 ± 4.35% infested plants). The Bt products Genuity Trecepta and Optimum Leptra containing Cry genes and the Vip3Aa20 gene had the lowest levels of larval infestation in every experiment with very little to no infestation at R3 growth stage in both years (Tables 2 and 3). Bt hybrids expressing only Cry proteins provided intermediate infestation control, but infestation levels varied in statistical significance from non-Bt hybrids. Genuity VT Double PRO (Cry1A.105 + Cry2Ab2) and SmartStax (Cry1A.105 + Cry2Ab2 + Cry1Fa2) reduced corn earworm infestations in 2019, but only SmartStax had reduced infestation in 2020. Infestations of Optimum Intrasect (Cry1Ab + Cry1Fa2) were not significantly different from those in comparable non-Bt hybrids.

The results for average corn earworm larvae per ear were similar to those of infestation levels. Hybrids exhibited significantly different numbers of total larvae per ear among hybrids in 2019 ($F = 48.82$; df = 9, 54; $p < 0.0001$) and 2020 ($F = 17.39$; df = 11, 66; $p < 0.0001$). The Pioneer brand hybrids as a group tended to have more larvae per ear than the Dekalb hybrids (Figures 1 and 2). A total of only five larvae and three exit holes were observed in plants of all hybrids expressing Vip3Aa20 in all experiments with almost all of these being in the early planting in 2020. Bt hybrids expressing only Cry proteins with the Cry2Ab2 protein had significantly fewer larvae than non-Bt hybrids in both years, whereas total larvae counts were similar for the hybrids expressing Cry1Ab + Cry1Fa2 and comparable non-Bt hybrids in both years (Figures 1 and 2). The number of larvae within each size category was significantly different for all four size categories in 2019 (F = 12.41 to 15.72; df = 9, 54; $p < 0.0001$) and 2020 (F = 4.11 to 6.18; df = 11, 66; $p < 0.001$). Larvae from hybrids expressing only Cry proteins were mostly small or medium in size as compared with larvae in ears of non-Bt hybrids that were mostly large or had already exited the ear (Figures 1 and 2; Tables S1 and S2).

Table 2. Effect of Bt traits on LS means ± SEM of percentage corn earworm-infested ears and damaged area by ear region per ear at growth stage R6 in 2019.

Brand and Hybrid [a]	Bt Traits	Infested Ears at R3 (%)	Damaged Ears at R6 (%)	Damage (cm^2) by Ear Region		
				Ear Tip	Kernels	Total
DKC 6694	None (RR2)	55.0 ± 5.2 c	57.5 ± 5.3 c	0.92 ± 0.14 e	1.15 ± 0.28 cd	2.07 ± 0.38 d
DKC 6697	VT Double PRO +	28.3 ± 4.7 e	35.8 ± 6.7 d	0.47 ± 0.10 f	0.48 ± 0.15 ef	0.96 ± 0.22 e
DKC 6629	Trecepta ++	0 f	0 e	0 g	0 f	0 f
DKC 6205	None (RR2)	65.8 ± 6.1 b	61.7 ± 5.6 c	1.39 ± 0.16 cd	1.32 ± 0.26 cd	2.72 ± 0.35 cd
DKC 6208	SmartStax +	41.7 ± 6.0 d	50.8 ± 4.2 c	1.15 ± 0.16 de	0.93 ± 0.19 de	2.08 ± 0.18 d
Pio 1637R	None (RR2)	84.2 ± 5.2 a	90.8 ± 3.3 a	2.37 ± 0.19 b	2.75 ± 0.57 b	5.11 ± 0.66 b
Pio 1637YHR	Intrasect +	80.8 ± 4.8 a	77.5 ± 7.0 b	1.59 ± 0.30 c	1.60 ± 0.33 c	3.19 ± 0.57 c
Pio 1637VYHR	Leptra ++	0 f	0 e	0 g	0 f	0 f
Pio 2088R	None (RR2)	87.5 ± 5.4 a	96.7 ± 2.2 a	3.25 ± 0.16 a	3.47 ± 0.51 a	6.72 ± 0.57 a
Pio 2089VYHR	Leptra ++	1.7 ± 1.7 f	0 e	0 g	0 f	0 f
$F > (P)$ (df = 9, 54)		96.98 (<0.0001)	78.72 (<0.0001)	28.74 (<0.0001)	26.49 (<0.0001)	55.84 (<0.0001)

LS means ± SEM within columns followed by the same letter are not significantly different (pair-wise t-tests of LSM, α = 0.05). [a] DKC for Dekalb hybrids; P for Pioneer brand hybrids. + Bt hybrid expressing only pyramided *Cry* proteins. ++ Bt-hybrid expressing pyramided *Cry* proteins with Vip3Aa20.

Table 3. Effect of Bt traits on LS means ± SEM of percentage corn earworm-infested ears and damaged area by ear region per ear at R6 growth stage in 2020.

Brand and Hybrid [a]	Bt Traits	Infested Ears at R3 (%)	Damaged Ears at R6 (%)	Damage (cm^2) by Ear Region		
				Ear Tip	Kernels	Total
DKC 6694	None (RR2)	42.5 ± 14.3 bc	38.3 ± 14.3 bc	0.67 ± 0.26 cd	0.74 ± 0.28 de	1.42 ± 0.54 de
DKC 6697	VT Double PRO +	39.2 ± 13.6 bc	28.3 ± 10.5 c	0.37 ± 0.14 d	0.42 ± 0.17 e	0.78 ± 0.31 e
DKC 6629	Trecepta ++	2.5 ± 2.5 d	0.8 ± 0.8 d	0.02 ± 0.02 e	0.01 ± 0.01 f	0.02 ± 0.02 f
DKC 6205	None (RR2)	56.7 ± 10.8 ab	55.8 ± 15.8 ab	1.01 ± 0.32 bc	1.48 ± 0.49 cd	2.49 ± 0.81 cd
DKC 6208	SmartStax+	36.8 ± 13.4 c	35.0 ± 9.9 c	0.64 ± 0.20 cd	0.83 ± 0.28 de	1.47 ± 0.48 de
DKC 6824	None (RR2)	55.0 ± 10.6 abc	60.8 ± 10.6 a	1.32 ± 0.28 ab	2.32 ± 0.44 b	3.64 ± 0.72 ab
DKC 6826	VT Double PRO +	37.5 ± 10.9 c	35.8 ± 12.3 c	0.65 ± 0.25 cd	1.02 ± 0.45 de	1.67 ± 0.69 de
DKC 6799	Trecepta ++	3.3 ± 3.3 d	0.8 ± 0.8 d	0.01 ± 0.01 e	0.02 ± 0.02 f	0.02 ± 0.02 f
Pio 2088R	None (RR2)	49.2 ± 11.2 abc	65.0 ± 8.1 a	1.25 ± 0.21 ab	3.20 ± 0.62 a	4.45 ± 0.81 a
Pio 2089VYHR	Leptra ++	0	4.2 ± 2.5 d	0.11 ± 0.07 e	0.05 ± 0.05 f	0.16 ± 0.12 f
Pio 1637/1870R	None (RR2)	63.3 ± 11.1 a	71.7 ± 9.2 a	1.48 ± 0.27 a	2.42 ± 0.45 b	3.90 ± 0.70 ab
Pio 1637/1870YHR	Intrasect +	54.2 ± 10.9 abc	54.2 ± 10.4 ab	1.12 ± 0.22 ab	2.19 ± 0.50 bc	3.31 ± 0.70 bc
$F > (P)$ (df = 11, 66)		11.00 (<0.0001)	14.93 (<0.0001)	13.72 (<0.0001)	16.61 (<0.0001)	16.69 (<0.0001)

LS means ± SEM within columns followed by the same letter are not significantly different (pair-wise t-tests of LSM, α = 0.05). [a] DKC for Dekalb hybrids; P for Pioneer brand hybrids. + Bt hybrid expressing only pyramided *Cry* proteins. ++ Bt-hybrid expressing pyramided *Cry* proteins with Vip3Aa20.

Results for R6 growth stage ear damage were also significantly different between hybrids (Tables 2 and 3). The proportion of damage in the tip and kernel regions was similar in 2019, but in 2020 much more kernel damage occurred relative to tip ear damage. Overall Bt hybrids reduced total corn earworm ear damage by an average of 75.0% in 2019 (non-Bt: 4.16 ± 0.32 cm^2; Bt: 1.04 ± 0.26 cm^2) and 66.7% in 2020 (non-Bt: 3.18 ± 0.30 cm^2; Bt: 1.06 ± 0.25 cm^2). Bt hybrids expressing Vip3Aa20 were undamaged in 2019 and exhibited only minimal ear damage, located almost entirely within the ear tip region, during the early planting in 2020 (Tables 2 and 3). Bt hybrids expressing only Cry proteins with the Cry2Ab2 protein provided moderate but significant reductions in both tip and kernel damage compared to non-Bt hybrids but were not as effective as hybrids expressing Vip3Aa20 (Tables 2 and 3). Hybrids with Cry1Ab + Cry1F reduced tip and kernel damage in 2019 but did not reduce damage in 2020.

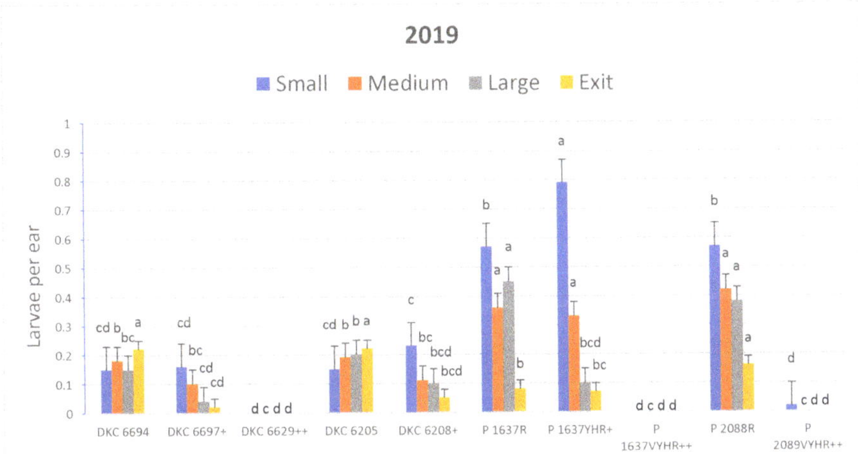

Figure 1. Effect of Bt traits on LS means ± SEM of number of corn earworm larvae per ear by size and exit holes in R3 growth stage 2019 field corn. LS means within larval size category with the same letter are not significantly different (pair-wise *t*-tests of LSM, α = 0.05). Bt hybrids marked with (+) expressed pyramided *Cry* proteins while hybrids marked with (++) expressed pyramided *Cry* proteins with Vip3Aa20. Unmarked hybrids were non-Bt hybrids. See Table 1 for specific proteins expressed for each entry and product type.

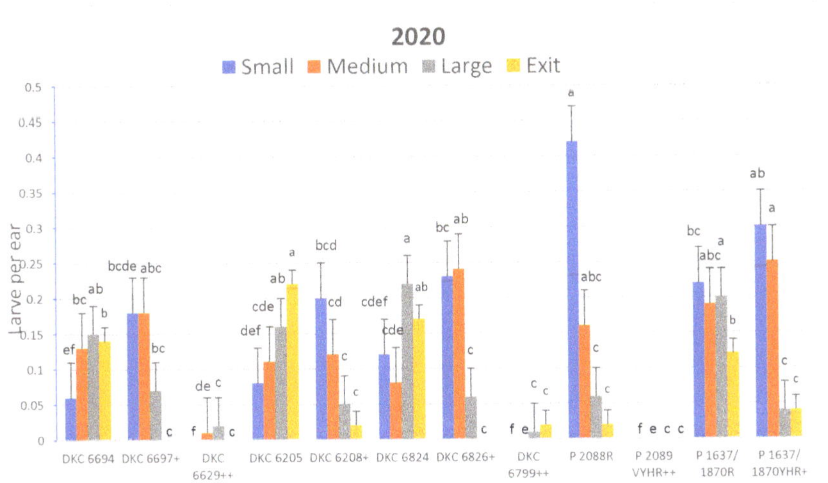

Figure 2. Effect of Bt traits on LS means ± SEM of number of corn earworm larvae per ear by size and exit holes in R3 growth stage 2020 field corn. LS means within larval size category with the same letter are not significantly different (pair-wise *t*-tests of LSM, α = 0.05). Bt hybrids marked with (+) expressed pyramided *Cry* proteins while hybrids marked with (++) expressed pyramided *Cry* proteins with Vip3Aa20. Unmarked hybrids were non-Bt hybrids. See Table 1 for specific proteins expressed for each entry and product type.

3.2. Grain Yield and Test Weight

Grain yields differed among hybrids in 2019, but these differences were not associated with Bt traits (F = 0.02; df = 1, 54; p = 0.8764) and instead reflected hybrid agronomics (Table 4). In 2020, grain yields were significantly different among hybrids and all Bt hybrids produced an average of 5.84% more grain yield than non-Bt hybrids (non-Bt: 12,642 ± 2270 kg/ha; Bt: 13,380 ± 1934 kg/ha; F = 11.95; df = 1, 66; p < 0.0009) (Table 5). Hybrids with Cry proteins and those with Vip3Aa20 both yielded more than the non-Bt hybrids (p = 0.0329 and p = 0.0044, respectively) but were not significantly different from each other (F = 2.82; df = 1, 66; p = 0.0975) (Table 5). Based on linear regression analysis, grain yield was not associated with corn earworm ear damage in 2019 (R^2 = 0.0061, F = 0.4791, p = 0.4909) but had a significantly negative association with corn earworm damage in 2020 (R^2 = 0.2638, F = 33.690, p < 0.0001) (Supplemental Figure S1).

Table 4. Effect of Bt traits on LS means ± SEM of corn grain yield, test weights, aflatoxin contamination, and fumonisin contamination in 2019.

Brand and Hybrid [a]	Bt Traits	Grain Yield (kg/ha)	Test Weight (kg/hL)	Aflatoxin (ppb)	Fumonisin (ppm)
DKC 6694	None (RR2)	14,732 ± 273 abcd	69.91 ± 1.67 a	57.00 ± 50.89 bc	32.00 ± 10.44 cde
DKC 6697	VT Double PRO [+]	13,954 ± 594 d	70.71 ± 1.62 a	39.81 ± 25.08 bc	31.75 ± 10.47 def
DKC 6629	Trecepta [++]	15,313 ± 651 abc	69.69 ± 1.74 a	3.31 ± 1.26 c	14.88 ± 6.07 f
DKC 6205	None (RR2)	15,658 ± 805 ab	70.11 ± 1.59 a	15.19 ± 10.96 bc	17.44 ± 6.44 ef
DKC 6208	SmartStax [+]	15,456 ± 779 ab	70.17 ± 1.57 a	29.19 ± 16.58 bc	17.25 ± 5.66 ef
Pio 1637R	None (RR2)	15,140 ± 404 abc	68.63 ± 2.70 ab	13.19 ± 8.89 bc	39.50 ± 9.75 bcd
Pio 1637YHR	Intrasect [+]	15,302 ± 477 abc	68.84 ± 2.06 ab	184.63 ± 56.43 a	57.13 ± 8.21 ab
Pio 1637VYHR	Leptra [++]	14,201 ± 593 cd	70.28 ± 1.77 a	16.56 ± 12.53 bc	25.44 ± 9.52 ef
Pio 2088R	None (RR2)	14,672 ± 682 bcd	63.31 ± 2.30 c	120.38 ± 75.79 b	75.13 ± 11.26 a
Pio 2089VYHR	Leptra [++]	15,830 ± 632 a	66.68 ± 2.55 b	116.98 ± 73.09 b	43.62 ± 9.15 abc
$F > (P)$ (df = 9, 54)		2.33 (0.0255)	7.10 (<0.0001)	3.15 (0.0036)	8.80 (<0.0001)

LS means ± SEM within columns followed by the same letter are not significantly different (pair-wise t-tests of LSM, α = 0.05). Statistical analyses based on $\log_{10}(X + 1)$ values. [a] DKC for Dekalb hybrids; P for Pioneer brand hybrids. [+] Bt hybrid expressing only pyramided *Cry* proteins. [++] Bt-hybrid expressing pyramided *Cry* proteins with Vip3Aa20.

Table 5. Effect of Bt traits on LS means ± SEM of corn grain yield, test weights, aflatoxin contamination, and fumonisin contamination in 2020.

Brand and Hybrid [a]	Bt Traits	Grain Yield (kg/ha)	Test Weight (kg/hL)	Aflatoxin (ppb)	Fumonisin (ppm)
DKC 6694	None (RR2)	12,029 ± 412 cd	75.04 ± 4.00 a	15.10 ± 14.99 bc	3.39 ± 0.87
DKC 6697	VT Double PRO [+]	13,770 ± 686 a	70.29 ± 0.40 abc	8.73 ± 7.77 bc	4.15 ± 0.91
DKC 6629	Trecepta [++]	13,478 ± 678 ab	70.00 ± 0.79 abc	0.01 ± 0.01 c	4.36 ± 0.76
DKC 6205	None (RR2)	12,498 ± 851 bcd	70.87 ± 1.78 abc	1.91 ± 1.59 bc	6.40 ± 2.65
DKC 6208	SmartStax [+]	13,272 ± 932 ab	72.45 ± 4.12 ab	25.46 ± 18.91 bc	6.99 ± 3.40
DKC 6824	None (RR2)	13,049 ± 751 abc	68.81 ± 1.05 bcd	1.37 ± 0.94 bc	3.14 ± 1.15
DKC 6826	VT Double PRO [+]	12,853 ± 652 abcd	68.91 ± 0.44 bcd	12.60 ± 12.49 bc	5.19 ± 2.18
DKC 6799	Trecepta [++]	13,706 ± 797 a	69.54 ± 0.89 bcd	0.91 ± 0.62 bc	3.35 ± 0.93
Pio 2088R	None (RR2)	13,739 ± 556 a	64.59 ± 0.82 d	85.01 ± 59.23 a	6.44 ± 1.44
Pio 2089VYHR	Leptra [++]	13,755 ± 727 a	66.24 ± 0.64 cd	58.26 ± 37.91 ab	5.63 ± 1.64
Pio 1637/1870R	None (RR2)	11,897 ± 1220 d	68.78 ± 1.47 bcd	0.29 ± 0.16 bc	4.27 ± 1.11
Pio 1637/1870YHR	Intrasect[+]	12,822 ± 376 abcd	70.19 ± 1.00 abc	53.14 ± 34.70 ab	3.50 ± 0.84
$F > (P)$ (df = 11, 66)		3.37 (0.0008)	2.13 (0.0263)	2.35 (0.0150)	1.08 (0.3906)

LS means ± SEM within columns followed by the same letter are not significantly different (pair-wise t-tests of LSM, α = 0.05). Statistical analyses based on $\log_{10}(X + 1)$ values. [a] DKC for Dekalb hybrids; P for Pioneer brand hybrids. [+] Bt hybrid expressing only pyramided *Cry* proteins. [++] Bt hybrid expressing pyramided *Cry* proteins with Vip3Aa20.

3.3. Mycotoxin Contamination Levels

Grain aflatoxin contamination in 2019 and 2020 was highly variable among plot samples. While most concentrations remained around the federal standard limit of aflatoxin in grain intended for human consumption of 20 ppb, a small number of samples exceeded 100 ppb with a few exceeding 500 ppb. Contamination levels were significantly different among hybrids, most likely due to genetic background, because aflatoxin levels were not different between hybrids with and without Bt traits in 2019 ($F = 0.57$; df = 1, 54; $p = 0.4524$) and 2020 ($F = 0.03$; df = 1, 66; $p = 0.8543$) (Tables 4 and 5). Variability in concentrations for a select few points resulted in skewing of statistical significance among hybrids. The Bt hybrid Pioneer 1637YHR, an Optimum Intrasect product, had an average aflatoxin contamination level significantly greater than those of all other hybrids in 2019 (Table 4). Linear regression analyses showed that aflatoxin contamination levels were not significantly associated with corn earworm damage in either year (2019: $R^2 = 0.0158$, $F = 1.2517$, $p = 0.2667$; 2020: $R^2 = 0.0249$, $F = 2.4050$, $p = 0.1243$) (Figure 3).

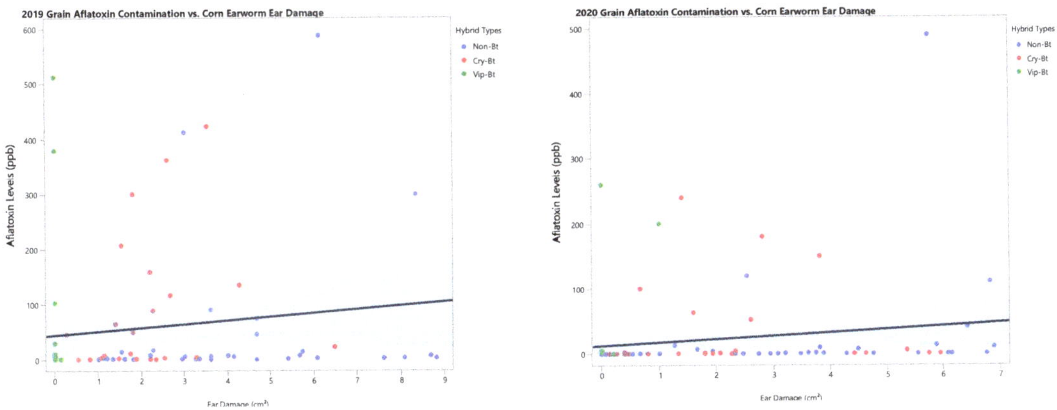

Figure 3. Linear regression analysis depicting the relationship between corn earworm ear damage and grain aflatoxin contamination levels of field corn by year. 2019: $R^2 = 0.0158$; $F = 1.2517$; $p = 0.2667$. 2020: $R^2 = 0.0249$; $F = 2.4050$; $p = 0.1243$.

Grain fumonisin contamination in 2019 was variable among plots but was significantly different among hybrids (Table 4). Only Bt hybrids expressing the Vip3Aa20 protein had significantly lower contamination levels compared to non-Bt hybrids ($F = 17.49$, df = 1, 54; $p < 0.0001$). However, grain fumonisin contamination levels for all hybrids in 2019 exceeded the federal standard limit of 2–4 ppm for grain intended for human consumption [39]. Bt hybrids expressing only Cry proteins did not have significantly reduced fumonisin contamination as compared to non-Bt hybrids ($F = 1.11$, df = 1, 54; $p = 0.2962$), and were not statistically different from Bt hybrids expressing Vip3Aa20 ($F = 1.93$, df = 1, 54; $p = 0.1707$). Overall contamination levels in 2020 were much lower than in 2019 and were not significantly different among hybrids (Table 5). Grain fumonisin contamination levels in 2020 also were not significantly different among Bt traits ($F = 0.02$, df = 1, 66; $p = 0.0900$). Grain fumonisin contamination levels had a significant positive association with corn earworm ear damage in both 2019 and 2020, with the association being stronger in 2019 when contamination levels were higher (Figure 4).

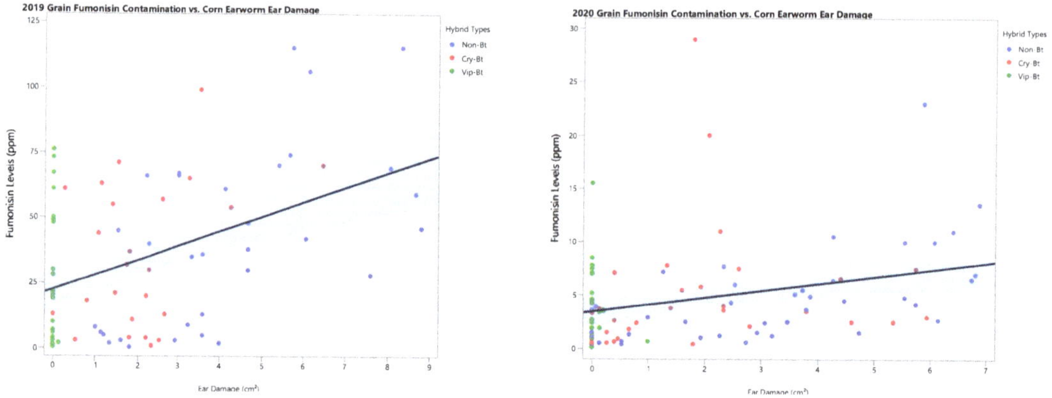

Figure 4. Linear regression analysis depicting the relationship between corn earworm ear damage and grain fumonisin contamination levels of field corn by year. 2019: $R^2 = 0.1970$; $F = 19.141$; $p < 0.0001$. 2020: $R^2 = 0.0909$; $F = 9.3961$; $p = 0.0028$.

4. Discussion

Planting transgenic corn hybrids that express multiple Bt traits through gene pyramiding has been a continuing protocol enacted in temperate and subtropical regions where lepidopteran pest infestation, including corn earworm, is common [14,16,49,61]. Resistance to specific Bt traits has been increasing in local corn earworm populations across the southeastern United States region, making them more difficult to control [17,19,20,42,62–65].

Bt corn hybrids expressing the Vip3Aa20 protein were the most effective in reducing corn earworm infestations and ear damage in all experiments. These results are consistent with other studies that show Bt products expressing Vip3Aa20 having minimal or no earworm infestations [19,21,64,65]. Bt hybrids expressing only Cry proteins provided moderate reductions in earworm infestation and damage. Despite having similar numbers of total corn earworm larvae in ears to those of non-Bt hybrids, Bt hybrids expressing only Cry proteins had mostly small and medium-sized larvae, while larvae from non-Bt hybrids were mostly large or were exiting the ears to pupate in the soil. This indicates that while the Cry protein did not cause much direct mortality, the toxins did delay the development of corn earworm larvae. It is unknown if any surviving corn earworms from these Bt hybrid plots had reduced adult survival and fecundity [66]. Bt hybrids with pyramided Cry proteins alone also caused partial reductions in both tip and kernel damage. Bt hybrids that expressed Cry1Ab + Cry1Fa2 proteins failed to prevent any ear infestation or damage caused by corn earworm larvae. Cry1Fa2 has not been considered an effective insecticidal source against corn earworm in the ear [67], and the potency of Cry1Ab has declined due to widespread resistance against the toxin in corn earworm populations across the southeastern United States region [18–20,63,64,66]. Partial earworm control at optimally timed plantings was most notable for Bt products expressing Cry1A.105 + Cry2Ab2, yet these products remain vulnerable to further resistance development [20,66,68]. These data suggest that Bt hybrids will continue to provide effective control of lepidopteran pest species in the southeastern U.S. including the European corn borer, fall armyworm, and corn earworm [7,8,48,66,67,69,70]. Future use of Cry proteins in regions with notable corn earworm activity must be carefully considered to extend and manage the efficacy of available Bt products.

While corn earworms are well-known for infesting ears and feeding on the kernels, the resulting injury normally does not qualify them as an economically significant pest of field corn [4,67,71]. Simulated corn earworm damage only resulted in significant yield loss when at least 60 kernels, or 15 cm^2 based on the conversion 0.25 cm^2 = 1 kernel, were damaged per ear [5]. Mean kernel damage in both 2019 and 2020 was much lower than the equivalent of 60 kernels per ear even in the late plantings. However, all Bt hybrids preserved significantly more grain yield than non-Bt hybrids in 2020, when ear damage in non-Bt hybrids was greatest and earworm damage was correlated with grain yield. While corn earworm currently may not be a primary issue in Georgia corn production, Bt hybrids can be effective in reducing kernel loss because of larval ear feeding [3,70] and may in some cases prevent grain yield loss, especially in later plantings [68]. Determining whether Bt hybrids could also improve grain quality based on test weight requires further evaluation.

Aflatoxin contamination accumulates from *Aspergillus* spp. infection and is notorious for emerging in an array of crops, including corn, whether transferred from pests [41] or developing through induced plant stress conditions including insect feeding injury [29,31,34,41,43]. Early studies suggest higher grain aflatoxin contamination levels occur in plants that have greater amounts of lepidopteran pest ear damage, especially by corn earworm [26,40]. Our data suggest that grain aflatoxin contamination was not associated with corn earworm ear damage nor was it reduced by the presence of any Bt toxins. Varying levels of fungal contamination were observed in similar studies that found a lack of association between corn earworm ear injury and grain aflatoxin levels [4,7,46,50]. Relative differences in aflatoxin contamination levels may instead be influenced by environmental factors such as high temperature and humidity, kernel moisture content, and rainfall across planting locations and dates [30,34,35]. Production systems that minimize plant stress conditions such as early planting and irrigation may assist with reducing grain aflatoxin contamination in subtropical regions.

Bt corn hybrids in this study effectively reduced corn earworm infestations and grain fumonisin contamination accumulated from *Fusarium* spp. infection based on linear regression analysis in both years. Past reports have found associations between fumonisin contamination with lepidopteran pest species including the European corn borer and western bean cutworm [27,48,51,52,72,73]. In a meta-analysis of 21 years of studies, Bt technology reduced fumonisin contamination in corn grain by 30.6% but this analysis included results for studies worldwide with a variety of target lepidopteran pest species [53]. Work specifically on the relationship between corn earworm damage and fumonisin contamination is limited. Bowers et al. [49] observed strong associations between increased fumonisin contamination and kernel injury resulting from three lepidopteran pest species that were all significantly reduced in corn expressing Cry1Ab + Vip3Aa20. Other studies also found an association with lepidopteran ear damage and fumonisin levels in subtropical locations, but they did not identify a specific lepidopteran species [32,36]. Here the authors provide supporting data on fumonisin contamination levels in corn hybrids expressing a wider selection of pyramided Bt proteins intended for lepidopteran pest control, which showed that Bt hybrids expressing Vip3Aa20 were associated with reduced grain fumonisin contamination levels. Nevertheless, overall contamination was still over the federal standard for grain intended for direct human consumption and only partially met the standard for certain animal feeds [39].

5. Conclusions

This study demonstrates the importance of Bt corn hybrids expressing the Vip3Aa20 protein for corn earworm control in the southeastern United States region. Planting transgenic Bt corn hybrids expressing Vip3Aa20 reduced grain fumonisin contamination but did not reduce aflatoxin contamination. Older Cry protein Bt products provided only limited control of corn earworm damage and did not consistently reduce mycotoxin contamination. Increased use of the Vip3Aa20 protein across more than one crop for controlling Cry-resistant *H. zea* populations could reduce its durability [74,75], thereby emphasizing the importance of insecticide resistance management tactics such as planting non-Bt refuges and careful management of pyramided Bt products to protect its durability [21,74–78]. Bt products containing Vip3Aa20 can be combined with improved corn genetics and agronomic practices that reduce crop stress to mitigate fumonisin contamination of corn grain.

Supplementary Materials: The following supporting information can be downloaded at https://www.mdpi.com/article/10.3390/insects15120914/s1, Table S1. Effect of Bt traits on LS means ± SEM of percentage corn earworm infested ears and number of corn earworm larvae per ear by size and exit holes in R3 growth stage in 2019. Table S2. Effect of Bt traits on LS means ± SEM of percentage corn earworm infested ears and number of corn earworm larvae per ear by size and exit holes in R3 growth stage in 2020. Figure S1. Linear regression analysis depicting the relationship between corn earworm ear damage and grain yield of field corn by year. 2019: $R^2 = 0.0061$; $F = 0.4791$; $p = 0.4909$. 2020: $R^2 = 0.2638$; $F = 33.690$; $p < 0.0001$.

Author Contributions: Conceptualization, G.D.B.; methodology, W.Y.B. and G.D.B.; software, G.D.B.; validation, W.Y.B. and G.D.B.; formal analysis, W.Y.B. and G.D.B.; investigation, W.Y.B.; resources, W.Y.B. and G.D.B.; data curation, W.Y.B. and G.D.B.; writing—original draft preparation, W.Y.B.; writing—review and editing, G.D.B. and M.D.T.; visualization, G.D.B. and M.D.T.; supervision, G.D.B.; project administration, G.D.B.; funding acquisition, G.D.B. All authors have read and agreed to the published version of the manuscript.

Funding: This research was funded in part by a grant from Bayer CropScience. We also thank Bayer CropScience and Pioneer Hi-Bred International for providing corn seed. This research was funded in part by Hatch project funds (accession no. 1009397) from the U.S. Department of Agriculture National Institute of Food and Agriculture.

Data Availability Statement: The available data is presented in the paper and supplemental tables and figures.

Acknowledgments: We thank Brett Byous, Suzanne Deeb, Anna Agi, Joshua Joiner, and Ben Poole for technical assistance. We thank Conor Fair for statistical advice.

Conflicts of Interest: The authors declare that this study received funding from Bayer CropScience. The funder was not involved in the study design, collection, analysis, interpretation of data, the writing of this article or the decision to submit it for publication.

References

1. Koziel, M.G.; Beland, G.L.; Bowman, C.; Carozzi, N.B.; Crenshaw, R.; Crossland, L.; Dawson, J.; Desai, N.; Hill, M.; Kadwell, S.; et al. Field performance of elite transgenic maize plants expressing an insecticidal protein derived from *Bacillus thuringiensis*. *Nat. Biotechnol.* **1993**, *11*, 194–200. Available online: https://www.nature.com/articles/nbt0293-194 (accessed on 22 April 2024). [CrossRef]
2. Ostlie, K.R.; Hutchison, W.D.; Hellmich, R.L. *Bt Corn and European Corn Borer*; North Central Regional Publication 602; Iowa State University Press: Ames, IA, USA, 1997.
3. Reay-Jones, F.P.; Reisig, D.D. Impact of corn earworm injury on yield of transgenic corn producing Bt toxins in the Carolinas. *J. Econ. Entomol.* **2014**, *107*, 1101–1109. [CrossRef] [PubMed]
4. Bibb, J.L.; Cook, D.; Catchot, A.; Musser, F.; Stewart, S.D.; Leonard, B.R.; Buntin, G.D.; Kerns, D.; Allen, T.W.; Gore, J. Impact of corn earworm (Lepidoptera: Noctuidae) on field corn (Poales: Poaceae) yield and grain quality. *J. Econ. Entomol.* **2018**, *111*, 1249–1255. [CrossRef] [PubMed]
5. Olivi, B.M.; Gore, J.; Musser, F.M.; Catchot, A.L.; Cook, D.R. Impact of simulated corn earworm (Lepidoptera: Noctuidae) kernel feeding on field corn yield. *J. Econ. Entomol.* **2019**, *112*, 2193–2198. [CrossRef] [PubMed]
6. Dicke, F.F.; Guthrie, W.D. The most important corn insects. In *Corn and Corn Improvement*, 3rd ed.; Sprague, G.F., Dudley, J.W., Eds.; American Society of Agronomy number 18; American Society of Agronomy: Madison, WI, USA, 1988; pp. 769–850.
7. Buntin, G.D.; Lee, R.D.; Wilson, D.M.; McPherson, R.M. Evaluation of YieldGard transgenic resistance for control of fall armyworm and corn earworm (Lepidoptera: Noctuidae) on corn. *Fla. Entomol.* **2001**, *84*, 37–42. [CrossRef]
8. Buntin, G.D.; All, J.N.; Lee, R.D.; Wilson, D.M. Plant-incorporated *Bacillus thuringiensis* resistance for control of fall armyworm and corn earworm (Lepidoptera: Noctuidae) in corn. *J. Econ. Entomol.* **2004**, *97*, 1603–1611. [CrossRef]
9. Cook, D.R.; Crow, W.; Gore, J.; Threet, M. Performance of selected Bt corn hybrids/technologies against corn earworm 1, 2020. *Arthropod Manag. Tests* **2021**, *46*, tsab087. [CrossRef]
10. Reay-Jones, F.P.F.; Buntin, G.D.; Reisig, D.D. Interactive effects between yields of Bt and non-Bt corn and planting dates in the southeastern United States. *J. Econ. Entomol.* submitted.
11. USDA-ERS (Economic Research Service). Recent Trends in GE Adoption. 2020. Available online: https://www.ers.usda.gov/data-products/adoption-of-genetically-engineered-crops-in-the-us/recent-trends-in-ge-adoption.aspx (accessed on 22 April 2024).
12. DeLamar, Z.D.; Flanders, K.L.; Holliman, J.L.; Mask, P.L. Efficacy of transgenic corn against southern insect pests in Marion Junction, Alabama, 1998. *Arthropod Manag. Tests* **1999**, *24*, M8. [CrossRef]

13. DeLamar, Z.D.; Flanders, K.L.; Pegues, M.D.; Mask, P.L. Efficacy of transgenic corn against southern insect pests in Fairhope, Alabama, 1998. *Arthropod Manag. Tests* **1999**, *24*, M9. [CrossRef]
14. Buntin, G.D.; Flanders, K.L.; Lynch, R.E. Assessment of experimental Bt events against fall armyworm and corn earworm in field corn. *J. Econ. Entomol.* **2004**, *97*, 259–264. [CrossRef] [PubMed]
15. Drury, S.M.; Reynolds, T.L.; Ridley, W.P.; Bogdanova, N.; Riordan, S.; Nemeth, M.A.; Sorbet, R.; Trujillo, W.A.; Breeze, M.L. Composition of forage and grain from second-generation insect-protected corn MON 89034 is equivalent to that of conventional corn (*Zea mays* L.). *J. Agric. Food Chem.* **2008**, *56*, 4623–4630. [CrossRef] [PubMed]
16. Rule, D.M.; Nolting, S.P.; Prasifka, P.L.; Storer, N.P.; Hopkins, B.W.; Scherder, E.F.; Siebert, M.W.; Hendrix, W.H., III. Efficacy of pyramided Bt proteins Cry1F, Cry1A.105, and Cry2Ab2 expressed in SmartStax corn hybrids against lepidopteran insect pests in the northern United States. *J. Econ. Entomol.* **2014**, *107*, 403–409. [CrossRef] [PubMed]
17. Kaur, G.; Guo, J.; Brown, S.; Head, J.P.; Price, P.A.; Paula-Moraes, S.; Ni, X.; Dimase, M.; Huang, F. Field-evolved resistance of *Helicoverpa zea* (Boddie) to transgenic maize expressing pyramided Cry1A.105/Cry2Ab2 proteins in northeast Louisiana, the United States. *J. Invertebr. Pathol.* **2019**, *163*, 11–20. [CrossRef] [PubMed]
18. Yang, F.; González, J.C.S.; Williams, J.; Cook, D.C.; Gilreath, R.T.; Kerns, D.L. Occurrence and ear damage of *Helicoverpa zea* on transgenic *Bacillus thuringiensis* maize in the field in Texas, U.S. and its susceptibility to Vip3A protein. *Toxins* **2019**, *11*, 102. [CrossRef]
19. Reay-Jones, F.P.; Bilbo, T.R.; Reisig, D.D. Decline in sublethal effects of Bt corn on corn earworm (Lepidoptera: Noctuidae) linked to increasing levels of resistance. *J. Econ. Entomol.* **2020**, *113*, 2241–2249. [CrossRef]
20. Dively, G.P.; Kuhar, T.P.; Taylor, S.; Doughty, H.B.; Holmstrom, K.; Gilrein, D.; Nault, B.A.; Ingerson-Mahar, J.; Whalen, J.; Reisig, D.; et al. Sweet corn sentinel monitoring for lepidopteran field-evolved resistance to Bt toxins. *J. Econ. Entomol.* **2021**, *114*, 307–319. [CrossRef]
21. Burkness, E.C.; Dively, G.; Patton, T.; Morey, A.C.; Hutchison, W.D. Novel Vip3A *Bacillus thuringiensis* (Bt) maize approaches high-dose efficacy against *Helicoverpa zea* (Lepidoptera: Noctuidae) under field conditions: Implications for resistance management. *GM Crops* **2010**, *1*, 337–343. [CrossRef]
22. Bilbo, T.R.; Reay-Jones, F.P.; Reisig, D.D.; Greene, J.K. Susceptibility of corn earworm (Lepidoptera: Noctuidae) to Cry1A.105 and Cry2Ab2 in North and South Carolina. *J. Econ. Entomol.* **2019**, *112*, 1845–1857. [CrossRef]
23. Cook, D.R.; Gore, J.; Crow, W. Performance of selected of Bt corn hybrids/technologies against corn earworm, 2016. *Arthropod Manag. Tests* **2020**, *45*, tsaa095. [CrossRef]
24. Yang, F.; González, J.C.S.; Little, N.; Reisig, D.D.; Payne, G.; Dos Santos, R.F.; Jurat-Fuentes, J.L.; Kurtz, R.; Kerns, D.L. First documentation of major Vip3Aa resistance alleles in field populations of *Helicoverpa zea* (Boddie) (Lepidoptera: Noctuidae) in Texas, USA. *Sci. Rep.* **2020**, *10*, 1–8. [CrossRef] [PubMed]
25. Yang, F.; Kerns, D.L.; Little, N.S.; González, J.C.S.; Tabashnik, B.E. Early warning of resistance to Bt toxin Vip3Aa in *Helicoverpa zea*. *Toxins* **2021**, *13*, 618. [CrossRef]
26. McMillian, W.W.; Wilson, D.M.; Widstrom, N.W. Aflatoxin contamination of preharvest corn in Georgia: A six-year study of insect damage and visible *Aspergillus flavus*. *J. Environ. Qual.* **1985**, *14*, 200–202. [CrossRef]
27. Abbas, H.K.; Zablotowicz, R.M.; Weaver, M.A.; Shier, W.T.; Bruns, H.A.; Bellaloui, N.; Accinelli, C.; Abel, C.A. Implications of Bt traits on mycotoxin contamination in maize: Overview and recent experimental results in southern United States. *J. Agric. Food Chem.* **2013**, *61*, 11759–11770. [CrossRef] [PubMed]
28. Munkvold, G.P. Cultural and genetic approaches to managing mycotoxins in maize ears. *Annu. Rev Phytopath.* **2003**, *17*, 233–239.
29. Pruter, L.S.; Weaver, M., Brewer, M.J. Overview of risk factors and strategies for management of insect-derived ear injury and aflatoxin accumulation for maize grown in subtropical areas of North America. *J. Integr. Pest Manag.* **2020**, *11*, 13. [CrossRef]
30. Abbas, H.K.; Wilkinson, J.R.; Zablotowicz, R.M.; Accinelli, C.; Abel, C.A.; Bruns, H.A.; Weaver, M.A. Ecology of *Aspergillus flavus*, regulation of aflatoxin production, and management strategies to reduce aflatoxin contamination of corn. *Toxin Rev.* **2009**, *28*, 142–153. [CrossRef]
31. Pruter, L.S.; Brewer, M.J.; Weaver, M.A.; Murray, S.C.; Isakeit, T.S.; Bernal, J.S. Association of insect-derived ear injury with yield and aflatoxin of maize hybrids varying in Bt transgenes. *Environ. Entomol.* **2019**, *48*, 1401–1411. [CrossRef] [PubMed]
32. Parsons, M.W.; Munkvold, G.P. Effects of planting date and environmental factors on fusarium ear rot symptoms and fumonisin B1 accumulation in maize grown in six North American locations. *Plant Path.* **2012**, *61*, 1130–1142. [CrossRef]
33. CDC (U.S. Centers for Disease Control). Outbreak of aflatoxin poisoning— Eastern and central provinces, Kenya, January-July 2004. *Morb. Mortal. Wkly. Rep.* **2004**, *53*, 790–793.
34. Smith, M.S.; Riley, T.J. Direct and interactive effects of planting date, irrigation, and corn earworm (Lepidoptera: Noctuidae) damage on aflatoxin production in preharvest field corn. *J. Econ. Entomol.* **1992**, *85*, 998–1006. [CrossRef]
35. Wiatrak, P.J.; Wright, D.L.; Marois, J.J.; Wilson, D. Influence of planting date on aflatoxin accumulation in Bt, non-Bt, and tropical non-Bt hybrids. *Agron. J.* **2005**, *97*, 440–445. [CrossRef]
36. Abbas, H.; Shier, W.; Cartwright, R. Effect of temperature, rainfall and planting date on aflatoxin and fumonisin contamination in commercial Bt and non-Bt corn hybrids in Arkansas. *Phytoprotection* **2007**, *88*, 41–50. [CrossRef]
37. Pruter, L.S.; Brewer, M.J.; Murray, S.C.; Isakeit, T.; Pekar, J.J.; Wahl, N.J. Yield, insect-derived ear injury, and aflatoxin among developmental and commercial maize hybrids adapted to the North American subtropics. *J. Econ. Entomol.* **2020**, *113*, 2950–2958. [CrossRef]

38. FDA. Guidance for Industry: Action Levels for Poisonous or Deleterious Substances in Human Food and Animal Feed. 2000. Available online: https://www.fda.gov/regulatory-information/search-fda-guidance-documents/guidance-industry-action-levels-poisonous-or-deleterious-substances-human-food-and-animal-feed (accessed on 22 April 2024).
39. FDA. Guidance for Industry: Fumonisin Levels in Human Foods and Animal Feeds. 2001. Available online: https://www.fda.gov/regulatory-information/search-fda-guidance-documents/guidance-industry-fumonisin-levels-human-foods-and-animal-feeds (accessed on 22 April 2024).
40. Lillehoj, E.B.; Kwolek, W.F.; Fennell, D.I.; Milburn, M.S. Aflatoxin incidence and association with bright greenish-yellow fluorescence and insect damage in a limited survey of freshly harvested high-moisture corn. *Cereal Chem.* **1975**, *52*, 403–411.
41. Abel, C.A.; Abbas, H.K.; Zablotowicz, R.M.; Pollan, M.; Dixon, K. The association between corn earworm damage and aflatoxin production in preharvest maize grain. In Proceedings of the 3rd Fungal Genomics, 4th Fumonisin Elimination and 16th Aflatoxin Elimination Workshop, Savannah, Georgia, 13–15 October 2003; Robens, J.F., Brown, R.L., Eds.; p. 47.
42. Windham, G.L.; Williams, P.W.; Davis, F.M. Effects of the southwestern corn borer on *Aspergillus flavus* kernel infection and aflatoxin accumulation in maize hybrids. *Plant Dis.* **1999**, *83*, 535–540. [CrossRef] [PubMed]
43. Williams, W.P.; Windham, G.L.; Buckley, P.M.; Daves, C.A. Aflatoxin accumulation in conventional and transgenic corn hybrids infested with southwestern corn borer (Lepidoptera: Crambidae). *J. Agric. Urban Entomol.* **2002**, *19*, 227–236.
44. Bruns, H.A.; Abbas, H.K. Planting date effects on Bt and non-Bt corn in the mid-South. *J. Agron.* **2006**, *98*, 100–106. [CrossRef]
45. Bowers, E.; Hellmich, R.; Munkvold, G. Comparison of fumonisin contamination using HPLC and ELISA methods in Bt and near-isogenic maize hybrids infested with European corn borer or western bean cutworm. *J. Agric. Food Chem.* **2014**, *62*, 6463–6472. [CrossRef]
46. Farias, C.A.; Brewer, M.J.; Anderson, D.J.; Odvody, G.N.; Xu, W.; Sétamou, M. Native maize resistance to corn earworm, *Helicoverpa zea*, and fall armyworm, *Spodoptera frugiperda*, with notes on aflatoxin content. *Southwest Entomol.* **2014**, *39*, 411–426.
47. Yu, J.; Hennessy, D.A.; Wu, F. The impact of Bt corn on aflatoxin-related insurance claims in the United States. *Sci. Rep.* **2020**, *10*, 10046. [CrossRef] [PubMed]
48. Munkvold, G.P.; Hellmich, R.L.; Rice, L.G. Comparison of fumonisin concentrations in kernels of transgenic Bt maize hybrids and nontransgenic hybrids. *Plant Dis.* **1999**, *83*, 130–138. [CrossRef] [PubMed]
49. Bowers, E.; Hellmich, R.; Munkvold, G. Vip3Aa and Cry1Ab proteins in maize reduce *Fusarium* ear rot and fumonisins by deterring kernel injury from multiple lepidopteran pests. *World Mycotoxin J.* **2013**, *6*, 127–135. [CrossRef]
50. Bowen, K.L.; Flanders, K.L.; Hagan, A.K.; Ortiz, B. Insect damage, aflatoxin content, and yield of Bt corn in Alabama. *J. Econ. Entomol.* **2014**, *107*, 1818–1827. [CrossRef]
51. Parker, N.S.; Anderson, N.R.; Richmond, D.S.; Long, E.Y.; Wise, K.A.; Krupke, C.H. Larval western bean cutworm feeding damage encourages the development of Gibberella ear rot on field corn. *Pest Manag. Sci.* **2016**, *73*, 546–553. [CrossRef] [PubMed]
52. Smith, J.L.; Limay-Rios, V.; Hooker, D.C.; Schaafsma, A.W. *Fusarium graminearum* mycotoxins in maize associated with *Striacosta albicosta* (Lepidoptera: Noctuidae) injury. *J. Econ. Entomol.* **2018**, *111*, 1227–1242. [CrossRef]
53. Pellegrino, E.; Bedini, S.; Nuti, M.; Ercoli, L. Impact of genetically engineered maize on agronomic, environmental and toxicological traits: A meta-analysis of 21 years of field data. *Sci. Rep.* **2018**, *8*, 3113. [CrossRef]
54. Davis, F.M.; Ng, S.S.; Williams, W.P. Visual rating scales for screening whorl-stage corn for resistance to fall armyworm. *Miss. Agric. For. Exp. Stn. Tech. Bull.* **1992**, *186*, 1–9.
55. Adendroth, L.J.; Elmore, R.W.; Boyer, M.J.; Marlay, S.K. *Corn Growth and Development*; PMR 109; Iowa States University Extension: Ames, IA, USA, 2011.
56. NEOGEN. Veratox®® for Aflatoxin. Available online: https://www.neogen.com/solutions/mycotoxins/veratox-aflatoxin/ (accessed on 10 August 2019).
57. NEOGEN. Veratox®® for Fumonisin. Available online: https://www.neogen.com/solutions/mycotoxins/veratox-fumonisin/ (accessed on 10 August 2019).
58. Latimer, G.W. *Official Methods of Analysis of AOAC International*, 21st ed.; Method 2001.06; AOAC International: Rockville, MD, USA, 2019. Available online: https://search.worldcat.org/title/Official-methods-of-analysis-of-AOAC-International/oclc/1112415371 (accessed on 10 September 2020).
59. SAS Institute. *SAS Version 9.4 User's Manual*; SAS Institute: Cary, NC, USA, 2013.
60. SAS Institute. *JMP Pro Version 15.0.0 User's Manual*; SAS Institute: Cary, NC, USA, 2019.
61. Siebert, M.W.; Nolting, S.P.; Hendrix, W.; Dhavala, S.; Craig, C.; Leonard, B.R.; Stewart, S.D.; All, J.N.; Musser, F.R.; Buntin, G.D.; et al. Evaluation of corn hybrids expressing Cry1F, Cry1A. 105, Cry2Ab2, Cry34Ab1/Cry35Ab1, and Cry3Bb1 against southern United States insect pests. *J. Econ. Entomol.* **2012**, *105*, 1825–1834. [CrossRef]
62. Huang, F.; Qureshi, J.A.; Meagher, R.L., Jr.; Reisig, D.D.; Head, G.P.; Andow, D.A.; Ni, X.; Kerns, D.; Buntin, G.D.; Niu, Y.; et al. Cry1F resistance in fall armyworm *Spodoptera frugiperda*: Single gene versus pyramided Bt maize. *PLoS ONE* **2014**, *9*, e112958. [CrossRef]
63. Niu, Y.; Meagher, R.L., Jr.; Yang, F.; Huang, F. Susceptibility of field populations of the fall armyworm (Lepidoptera: Noctuidae) from Florida and Puerto Rico to purified Cry1F and corn leaf tissue containing single and pyramided Bt genes. *Fla. Entomol.* **2013**, *96*, 701–713. [CrossRef]

64. Niu, Y.; Oyediran, I.; Yu, W.; Lin, S.; Dimase, M.; Brown, S.; Reay-Jones, F.P.F.; Cook, D.; Reisig, D.D.; Thrash, B.; et al. Populations of *Helicoverpa zea* (Boddie) in the southeastern United States are commonly resistant to Cry1Ab, but still susceptible to Vip3Aa20 expressed in MIR 162 corn. *Toxins* **2021**, *13*, 63. [CrossRef] [PubMed]
65. Welch, K.L.; Unnithan, G.C.; Degain, B.A.; Wei, J.; Zhang, J.; Li, X.; Tabashnik, B.E.; Carrière, Y. Cross-resistance to toxins used in pyramided Bt crops and resistance to Bt sprays in *Helicoverpa zea*. *J. Invertebr. Pathol.* **2015**, *132*, 149–156. [CrossRef] [PubMed]
66. Bilbo, T.R.; Reay-Jones, F.P.; Reisig, D.D.; Musser, F.R.; Greene, J.K. Effects of Bt corn on the development and fecundity of corn earworm (Lepidoptera: Noctuidae). *J. Econ. Entomol.* **2018**, *111*, 2233–2241. [CrossRef] [PubMed]
67. Buntin, G.D. Corn expressing Cry1Ab or Cry1F endotoxin for fall armyworm and corn earworm (Lepidoptera: Noctuidae) management in field corn for grain production. *Fla. Entomol.* **2008**, *91*, 523–530.
68. Reay-Jones, F.P.F.; Buntin, G.D.; Reisig, D.D.; Bridges, W.C., Jr. Longitudinal trails illustrate interactive effects between declining Bt efficacy against *Helicoverpa zea* and planting dates in corn. *J. Econ. Entomol.* **2024**, *117*, 1901–1912. [CrossRef]
69. Reisig, D.D.; Akin, D.S.; All, J.N.; Bessin, R.T.; Brewer, M.J.; Buntin, G.D.; Catchot, A.L.; Cook, D.; Flanders, K.L.; Huang, F.-N.; et al. Lepidopteran injury to corn containing single and pyramided Bt traits, and a refuge blend with Cry1Ab x Cry1F, in the southern United States. *J. Econ. Entomol.* **2015**, *108*, 157–165. [CrossRef]
70. Reay-Jones, F.P.F.; Bessin, R.T.; Brewer, M.J.; Buntin, G.D.; Catchot, A.L.; Cook, D.R.; Flanders, K.L.; Kerns, D.L.; Porter, R.P.; Reisig, D.D.; et al. Impact of Lepidoptera (Crambidae, Noctuidae, and Pyralidae) pests on corn containing pyramided Bt traits and a blended refuge in the southern United States. *J. Econ. Entomol.* **2016**, *109*, 1859–1871. [CrossRef]
71. Reay-Jones, F.P. Pest status and management of corn earworm (Lepidoptera: Noctuidae) in field corn in the United States. *J. Integr. Pest Manag.* **2019**, *10*, 19. [CrossRef]
72. Dowd, P.F. Biotic and abiotic factors limiting the efficacy of Bt corn in indirectly reducing mycotoxin levels in commercial fields. *J. Econ. Entomol.* **2001**, *94*, 1067–1074. [CrossRef]
73. Bakan, B.; Melcion, D.; Richard-Molard, D.; Cahagnier, B. Fungal growth and Fusarium mycotoxin content in isogenic traditional maize and genetically modified maize grown in France and Spain. *J. Agric. Food Chem.* **2002**, *50*, 728–731. [CrossRef] [PubMed]
74. Caprio, M.A.; Kurtz, R.; Catchot, A.; Kerns, D.; Reisig, D.; Gore, J.; Reay-Jones, F.P.F. The corn-cotton agroecosystem in the mid-southern United States: What insecticidal event pyramids should be used in each to extend Vip3A durability. *J. Econ. Entomol.* **2019**, *112*, 2894–2906. [CrossRef] [PubMed]
75. Tabashnik, B.E.; Fabrick, J.A.; Carriere, Y. Global patterns of insect resistance to transgenic Bt crops: The first 25 years. *J. Econ. Entomol.* **2023**, *116*, 297–309. [CrossRef]
76. Head, G.P.; Greenplate, J. The design and implementation of inset resistance management program for Bt crops. *GM Crops Food* **2012**, *3*, 144–153. [CrossRef] [PubMed]
77. Carrière, Y.; Fabrick, J.A.; Tabashnik, B.E. Can pyramids and seed mixtures delay resistance to Bt crops? *Trends Biotechnol.* **2016**, *34*, 291–302. [CrossRef]
78. Reisig, D.D.; DiFonzo, C.; Dively, G.; Farhan, Y.; Gore, J.; Smith, J. Best management practices to delay the evolution of Bt resistance in lepidopteran pests without high susceptibility to Bt toxins in North America. *J. Econ. Entomol.* **2022**, *115*, 16–26. [CrossRef]

insects

MDPI

Article

Impact of Temperature Variation on the Biological Traits and Lifecycle of *Spodoptera exigua* (Lepidoptera: Noctuidae): A Meta-Analysis Approach

Honghua Zhang †, Danping Xu †, Xingqi Deng, Zhiqian Liu, Zhipeng He, Junhao Wu and Zhihang Zhuo *

College of Life Science, China West Normal University, Nanchong 637002, China;
honghua_zhang@foxmail.com (H.Z.); xudanping@cwnu.edu.cn (D.X.); deng.xinqi@foxmail.com (X.D.); qnhtvxhp319123@foxmail.com (Z.L.); zhipeng_hh@foxmail.com (Z.H.); wujunhao0824@gmail.com (J.W.)
* Correspondence: zhuozhihang@foxmail.com; Tel.: +86-1311-197-3927
† These authors contributed equally to this work.

Simple Summary: *Spodoptera exigua* is a significant pest of crops, but its ability to adapt to new climates remains poorly understood. This study investigates how temperature, light cycle, and humidity affect the beet armyworm's lifecycle. By analyzing 264 data points from 33 published studies, we found that warmer temperatures, particularly above 20 °C, significantly enhance the beet armyworm's physiological functions. As temperatures increase, the developmental stages shorten, egg-laying decreases, and the pupal stage shortens, which leads to a longer adult lifespan. The research determined the ideal environmental factors for each developmental stage of the beet armyworm, providing crucial insights into its adaptability in changing climates. These findings are important for predicting beet armyworm population dynamics and developing better pest management strategies.

Abstract: *Spodoptera exigua* is a pest of considerable economic importance; however, detailed research into its ecological adaptability in newly invaded habitats is limited. This research performed a comprehensive analysis of the life history characteristics of *S. exigua* under varying temperature, photoperiod, and humidity conditions. A total of 264 studies that met the inclusion criteria were included in the analysis, and the data were examined using random-effects model, fixed-effects model, and meta-regression analysis techniques. The findings reveal that when temperatures exceed 20 °C, several biological parameters of *S. exigua* significantly increase, with the highest biological activity observed at 33 °C. As temperature rises, the duration of each developmental phase significantly decreases, accompanied by a reduction in the average number of eggs produced by females and a shorter pre-oviposition period. In addition, the pupal development period is shortened, resulting in a longer adult lifespan. By considering environmental variables such as temperature, photoperiod, and relative humidity, we identified the optimal conditions for the survival of each developmental stage of *S. exigua*. These results provide a foundation for predicting the population dynamics of this pest and contribute to the development of more effective pest control strategies.

Keywords: *Spodoptera exigua*; climate impact; pest management strategies; invasive species; meta-analysis; survival analysis

1. Introduction

The increase in worldwide temperatures affects insects in numerous ways, including their survival, reproductive patterns, migration behaviors, and geographical

https://doi.org/10.3390/insects16020155

distribution [1,2]. Climatic warming is also having negative effects on many insect species, leading to decreased fertility and increased mortality in overwintering species due to their poor adaptability to rapid climatic changes. In regions with warmer climates, insects experience faster developmental rates, leading to quicker population growth. This accelerated growth may contribute to their migration towards higher latitudes or altitudes [3]. Moreover, increasing temperatures enhance the reproduction and spread of specific pests, which intensifies the risk they pose to crops [4,5]. As a result, climate change could influence insect population patterns, especially by altering the distribution and frequency of pest outbreaks, which has significant consequences for both ecosystems and agriculture [6].

Spodoptera exigua (Lepidoptera: Noctuidae) is an extensively polyphagous pest that represents a major threat to numerous economically valuable crops in tropical and subtropical regions. This pest feeds on 170 plant species belonging to 35 different families, causing considerable damage [7]. Before the 1980s, *S. exigua* was mainly found in Beijing, Hebei, Henan, Shandong, and the Guanzhong area of Shaanxi Province. While it was also present in the Yangtze River Basin, as well as the Northeast and Northwest regions, its impact was relatively limited. However, by the late 1980s, the spread of *S. exigua* intensified, leading to more significant damage [8]. Along with damaging staple crops like corn, sorghum, and soybeans, *S. exigua* also causes considerable harm to economically important crops, such as vegetables, cotton, and sugar beets [9–11].

Climate change significantly affects the lifecycle, reproductive rate, distribution, behavior, and resistance of *S. exigua* to pesticides. Elevated temperatures could shorten its lifecycle, boost population density, and enable its spread into new regions, making pest management more challenging [12]. As a result, integrated pest management approaches are crucial, encompassing the optimization of agricultural planting patterns, strengthening biological control methods, and implementing accurate monitoring systems to address the challenges posed by climate change. Evaluating the effects of temperature fluctuations on *S. exigua* is vital for formulating effective control strategies.

Recent studies have concentrated on the biological traits, ecological aspects, artificial rearing, and breeding methods, as well as the migration behaviors of *S. exigua* [13–15]. While some research has explored the influence of temperature on *S. exigua* populations, most studies have focused on its effects on particular aspects of the pest's biology [16,17]. This research employs meta-analysis to assess, in quantitative terms, the wider effects of temperature variations on *S. exigua* [18]. Meta-analysis is a structured research approach designed to combine and synthesize findings from several independent studies, allowing for more comprehensive and reliable conclusions [19]. Using meta-analysis, this study highlights the sensitivity of various biological traits of *S. exigua* to temperature variations under different conditions, providing valuable insights for the development of effective pest management strategies.

2. Materials and Methods

2.1. Literature Search and Selection Process

To comprehensively assess the impact of global climate change on the biological traits of *S. exigua*, this research performed a review of relevant literature from multiple databases. The search was carried out from September to October 2024, mainly using databases such as Web of Science, PubMed, Scopus, and CNKI. Furthermore, pertinent review papers were manually reviewed to identify studies that were not included in the database searches. The terms used in the search included "Beet armyworm" (also known as "Lesser armyworm" or "Spiny bollworm"), "climate change" (alternatively referred to as "global warming"), "temperature", "precipitation", "photoperiod", and "biological characteristics" (including "life history traits", "development", "reproduction", etc.).

The process of literature screening was divided into two stages: initially, studies unrelated to *S. exigua* or climate change were excluded based on their titles and abstracts; secondly, the full text of the remaining studies was reviewed to further exclude those that did not satisfy the inclusion criteria. The criteria for inclusion were as follows: (1) the study investigates how temperature variations influence the development time, lifespan, lifecycle, egg-laying period, fertility, and hatching rate of *S. exigua*. (2) The study includes data on additional variables, such as relative humidity, photoperiod, or other environmental factors. (3) The study provides experimental data, including sample size, mean values, standard errors, or standard deviations.

2.2. Data Extraction

Data pertinent to the study were collected from research articles that fulfilled the inclusion criteria. The main variables extracted were temperature (indicating various experimental temperature levels in °C), relative humidity, and biological traits (including developmental rate, generation time, and reproductive capacity, such as oviposition). When crucial statistical data, such as standard deviation or sample size, was not explicitly reported, the information was extracted from graphs using tools like WebPlotDigitizer (Version 4.7). For experiments involving multiple variables or datasets with several treatment groups, the results from each treatment were considered as independent effect sizes.

2.3. Statistical Analysis

In this research, the rma.mv function available in the "metafor" package of R version 4.3 was employed to perform the subsequent analyses [20]. Initially, we utilized a random-effects model to compute the relative risk (RR+) and we assessed the heterogeneity measure I^2 across studies using restricted maximum likelihood (REML) estimation. Depending on the value of I^2, explanatory variables that could affect the heterogeneity of effect sizes were then incorporated [21]. Next, a random-effects model was applied to assess the overall effect size across all treatment groups. Finally, statistical tests were performed to assess the mean effect size and the 95% confidence interval (CI), and the heterogeneity indices I^2 and the Q statistic (Qt) [22]. We performed separate meta-analyses for different independent variables to evaluate the response of *S. exigua* at various developmental stages to temperature, assessing the strength of these temperature effects.

In this meta-analysis, both random-effects and fixed-effects models were applied to integrate findings from multiple studies. Because of the significant heterogeneity observed across studies (e.g., variations in experimental locations, treatment procedures, etc.), the random-effects model was considered a more appropriate choice to address this variability [23]. We considered climate change-related factors (e.g., temperature and relative humidity) as independent variables and the biological traits of *S. exigua* as dependent variables. Effect sizes were computed by utilizing the Log Response Ratio (LRR), which compared the biological traits between the climate factor treatment groups and the control groups. Statistical significance was evaluated using 95% confidence intervals. To assess how different climate factors affect the biological traits of *S. exigua*, we performed subgroup analyses for temperature and relative humidity separately.

The Q statistic and I^2 index were used to assess the heterogeneity among the studies. The Q statistic tests for the presence of heterogeneity, while the I^2 value quantifies the degree of variability. An increased I^2 value signifies a higher degree of variability among the studies. The heterogeneity statistic is calculated by testing the weighted sum of squared deviations with $k - 1$ degrees of freedom, providing a measure of variability across studies. If the 95% confidence interval for the effect size includes zero, it suggests no significant difference between the experimental and control groups ($p > 0.05$). If the 95% confidence interval is entirely positive, it suggests a notably larger effect size in the experimental group than in the control group ($p < 0.05$). On the other hand, if the entire 95% confidence interval falls below zero, it suggests a significantly smaller effect size in the experimental group relative to the control group ($p < 0.05$) [24]. The decision to include explanatory variables was based on the significance of the cumulative effect size relative to zero and the p-value of the Qt statistic. The potential explanatory variables taken into account were the impacts of humidity and photoperiod on the overall effect size. Additionally, temperature data were considered as a continuous variable to assess their impact on the mean effect size. In the meta-analysis, the total heterogeneity was divided into between-group heterogeneity (variance attributed to categorical factors) and within-group heterogeneity (residual variance), with significance evaluated using a $k - 1$ test [25]. We also assessed potential publication bias by utilizing funnel plots and conducting Egger's regression test [26]. If significant bias was detected, the "trim and fill" method was applied to correct for it. To assess publication bias, we analyzed the connection between effect size and sample size through the use of funnel plots. The presence of publication bias was determined based on the significance of the p-value [27]. If statistical significance persists after adjustment, this suggests that the findings are consistent and not affected by publication bias.

3. Results

3.1. Statistical Data

This research incorporates data from 33 different publications, encompassing a total of 264 individual observations. It examines 13 dependent variables related to temperature variation in *S. exigua*, including first-instar larvae ($n = 20$), second-instar larvae ($n = 20$), third-instar larvae ($n = 20$), fourth-instar larvae ($n = 20$), fifth-instar larvae ($n = 20$), sixth-instar larvae ($n = 5$), adult lifespan ($n = 30$), egg production ($n = 38$), larval period ($n = 27$), generation time ($n = 11$), pre-oviposition period ($n = 5$), average number of eggs per female ($n = 16$), and pupal period ($n = 52$) (Table 1).

Table 1. Dataset on biological indicators of *S. exigua* (TR stands for temperature range, CT stands for controlled temperature, N represents sample size, and V stands for variable.).

TR	CT	N	V
15.5–38 °C	15.5	20	First instar
15.5–38 °C	15.5	20	Second instar
15.5–38 °C	15.5	20	Third instar
15.5–38 °C	15.5	20	Fourth instar
15.5–38 °C	15.5	20	Fifth instar
20–36 °C	20	5	Sixth instar
20–36 °C	20	30	Adult longevity
15.5–40 °C	15.5	19	Egg
20–36 °C	20	10	Generation cycle
22–28 °C	22	27	Larval stage
15–40 °C	15	16	Mean eggs number of per female
20–36 °C	20	5	Pre-oviposition
15.5–38 °C	15.5	52	Pupa stage
15–40 °C	15.5	264	*S. exigua*

3.2. Comprehensive Meta-Analyses with Random-Effects and Fixed-Effects Models

The findings indicate that elevated temperatures improved the adaptability of *S. exigua*, with a combined mean effect size of −1.0733 (95% CI: −1.1663, −0.9803; Figure 1). As temperatures increased, all dependent variables related to *S. exigua* significantly decreased, except for the average number of eggs per female and the pre-oviposition period, which remained unchanged (Figure 2A). When temperature was considered as a continuous variable, changes in *S. exigua* were observed across various temperature gradients (Figure 2B). Biological indicators of *S. exigua* significantly increased once temperatures exceeded 15 °C (Figure 3A). Biological activity peaked at 33 °C (Figure 3A), with a photoperiod of 12:12 and humidity at 80% (Figure 3B).

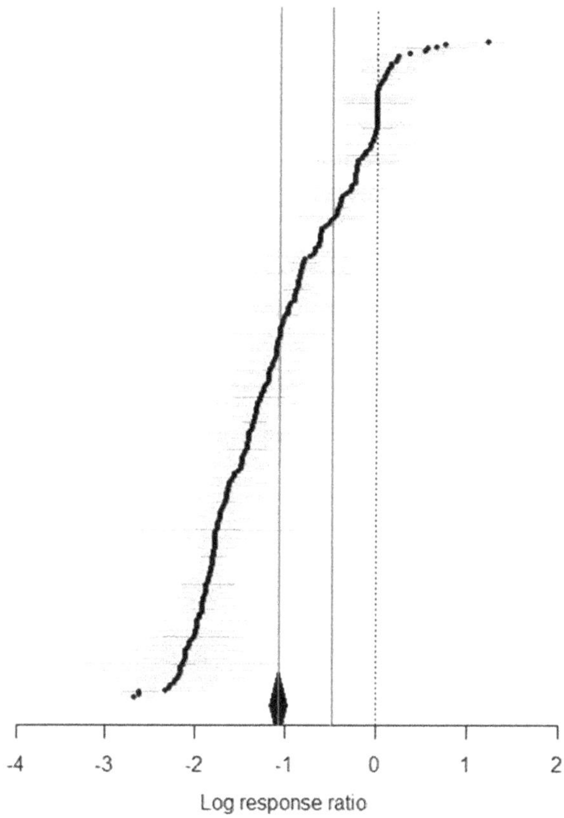

Figure 1. The forest plot shows the impact of temperature variation on *S. exigua*. The red line represents the results from the random-effects model, with a pooled effect size of −1.0733 and a 95% confidence interval from −1.1663 to −0.9803. The blue solid line indicates the results from the fixed-effects model, with a pooled effect size of −0.4942 and a 95% confidence interval ranging from −0.4985 to −0.4898. The black dashed line marks x = 0 and the gray line represents the standard error of individual factors.

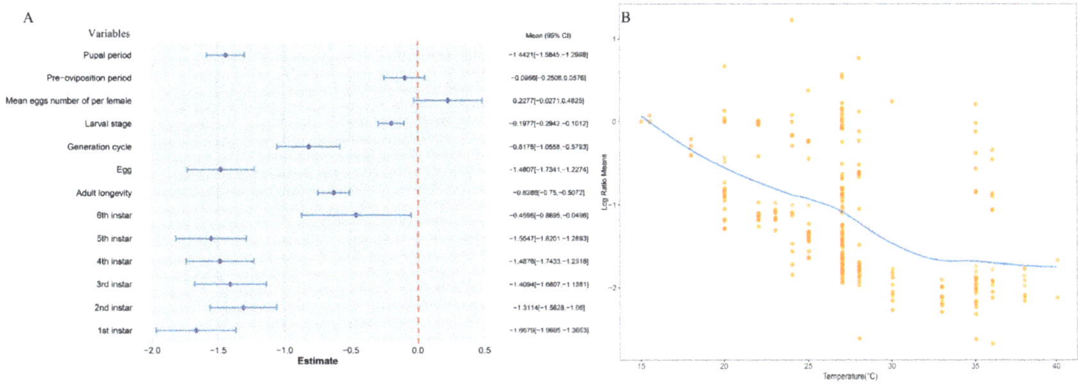

Figure 2. The impact of temperature variation on *S. exigua*. (**A**) illustrates the changes in biological traits of *S. exigua* as temperature increases. The red dashed line indicates x = 0. Dark blue squares represent the cumulative effect size for each temperature gradient, while the light blue line shows the 95% confidence interval. (**B**) depicts the temperature range curve for the optimal growth of *S. exigua*. The orange dots represents all factors and the blue solid line indicates the suitability of *S. exigua* to temperature variation. The shaded area indicates the 95% confidence interval.

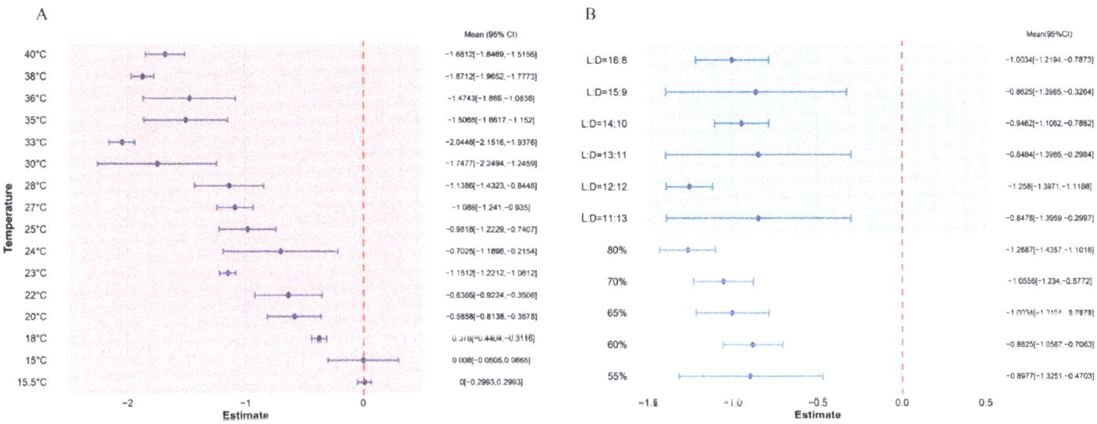

Figure 3. Forest plot showing the effects of temperature, photoperiod, and humidity on different biological indices of *S. exigua*. (**A**) illustrates how the physiological traits of *S. exigua* vary as the temperature increases. (**B**) depicts the response of *S. exigua* to changes in external environmental conditions. The dark blue squares indicate the cumulative effect size for each group, with the red dashed line representing x = 0. The light blue line marks the 95% confidence interval.

3.3. The Effect of Temperature on Developmental Duration

In all the studies examined, higher temperatures were associated with a decrease in the developmental period of first-instar *S. exigua*, resulting in a combined mean effect size of −1.6678 (CI: −1.9695, −1.3663; Figure S1). Within the temperature range of 15.5–38 °C, the developmental time of first-instar *S. exigua* decreased significantly as temperature increased (Figure 4A). Both the random-effects and fixed-effects models showed Q (df = 19) = 462.6090, $p < 0.0001$; this suggests that the variability across studies significantly influenced the cumulative effect size, highlighting the need to include explanatory variables (Figure 4B). The findings revealed that humidity (Qm = 5.1338, $p = 0.0235$) influences the cumulative effect size and that variations in relative humidity and photoperiod have distinct effects on the relationship between temperature variation and the developmental period of first-instar *S. exigua*. The optimal conditions for first-instar development were identified as 36 °C, 65% relative humidity, and a photoperiod of 16:8.

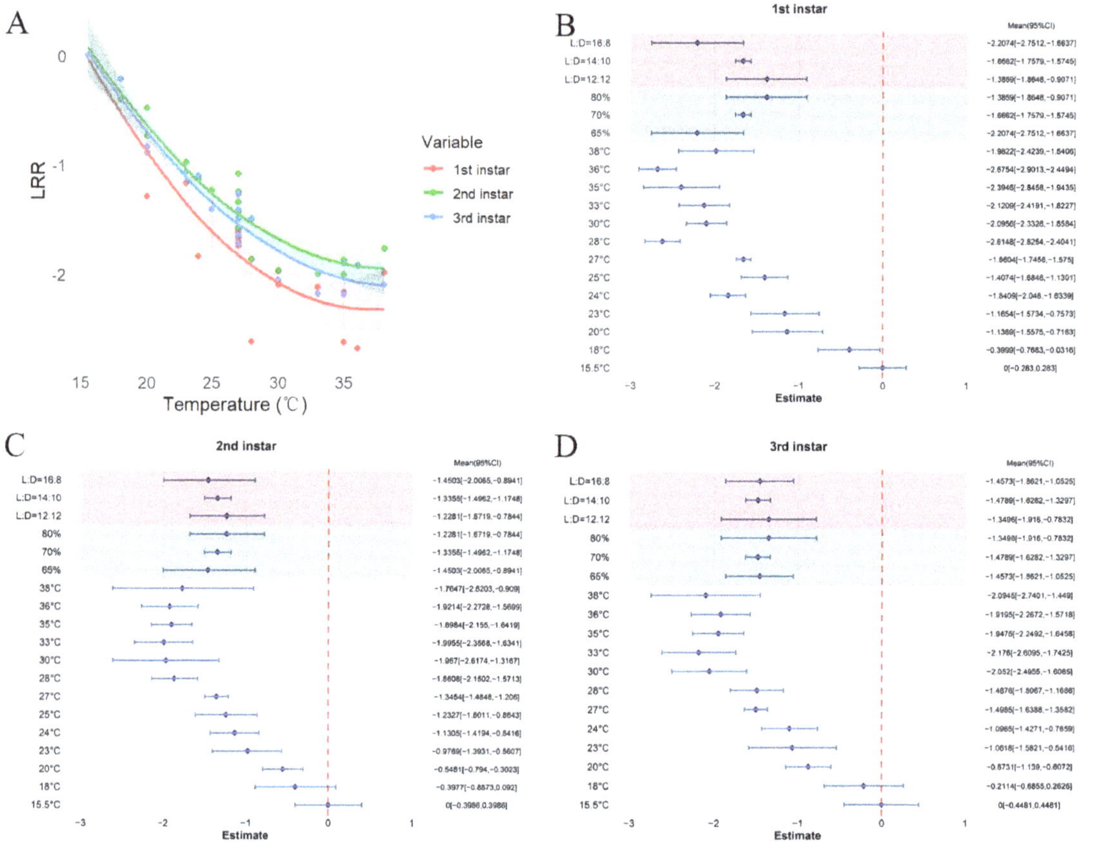

Figure 4. The impact of temperature on the developmental stages of *S. exigua*. (**A**) displays the variations in developmental time for first to third instar larvae at different temperature levels. (**B–D**) depict the reaction of first to third instar larvae to changes in external environmental conditions.

In all experiments, elevated temperatures led to a reduced developmental period for second-instar *S. exigua*, the pooled mean effect size was −1.3114 (CI: −1.5628, −1.0600; Figure S2). The developmental period of second-instar *S. exigua* decreased significantly with increasing temperature within the 15.5–38 °C range (Figure 4A). The findings from both random-effects and fixed-effects models indicated a significant between-study heterogeneity (Q (df = 19) = 167.9342, $p < 0.0001$), influencing the cumulative effect size, which underscores the necessity of incorporating explanatory variables (Figure 4C). These findings suggest that the optimal conditions for second-instar *S. exigua* development are 33 °C, 65% relative humidity, and a photoperiod of 16:8.

Across all studies, elevated temperatures resulted in a shortened developmental period for third-instar *S. exigua*, with a pooled mean effect size of −1.4094 (CI: −1.6807, −1.1381; Figure S3). Within the temperature range of 15.5–38 °C, the developmental time of third-instar *S. exigua* decreased significantly as temperature increased (Figure 4A). Results from both random-effects and fixed-effects models showed Q (df = 18) = 134.8382, $p <$ 0.0001; this suggests significant variability between studies that affected the cumulative effect size, emphasizing the importance of incorporating explanatory variables (Figure 4C). These findings suggest that the optimal conditions for third-instar *S. exigua* development are 33 °C, 70% relative humidity, and a photoperiod of 14:10.

Across the studies, elevated temperatures resulted in a shorter developmental period for fourth-instar *S. exigua*, with a combined mean effect size of −1.4876 (CI: −1.7433, −1.2318; Figure S4). Within the temperature span of 15.5–38 °C, the developmental period of fourth-instar *S. exigua* significantly shortened with rising temperatures (Figure 5A). Both random-effects and fixed-effects model results revealed Q (df = 19) = 192.8081, $p < 0.0001$, suggesting substantial heterogeneity between studies that affected the cumulative effect size, emphasizing the need for explanatory variables to account for this variability (Figure 5B). These findings suggest that the optimal conditions for fourth-instar *S. exigua* development are 33 °C, 65% relative humidity, and a photoperiod of 16:8.

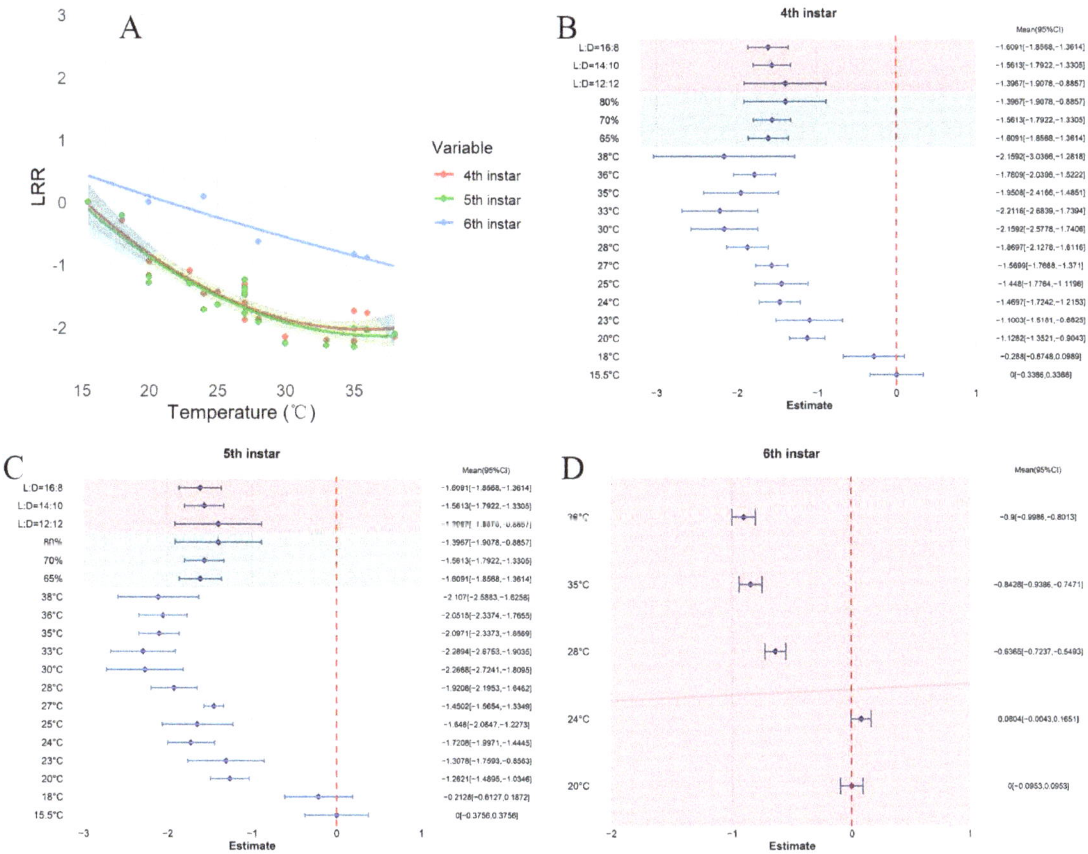

Figure 5. The impact of temperature on the developmental stages of *S. exigua*. (**A**) illustrates how developmental time for fourth to sixth instar larvae changes across varying temperatures. (**B–D**) represent the responses of fourth to sixth instar larvae to different external environmental conditions.

Across all studies, elevated temperatures led to a reduction in the developmental period of fifth-instar *S. exigua*, with a combined mean effect size of −1.5547 (CI: −1.8201, −1.2893; Figure S5). Within the 15.5–38 °C temperature range, the developmental time of fifth-instar *S. exigua* significantly decreased as the temperature rose (Figure 5A). The results from both random-effects and fixed-effects models indicated Q (df = 19) = 195.3696, $p < 0.0001$, highlighting considerable variability between studies that impacted the overall effect size, thus requiring the incorporation of explanatory variables (Figure 5C). The results indicate that both humidity and photoperiod have an impact on the overall effect size, with variations in these environmental factors affecting the developmental period of fifth-instar *S. exigua* in distinct ways. The optimal conditions for fifth-instar development were identified as 33 °C, 65% relative humidity, and a photoperiod of 16:8.

Across all studies, elevated temperatures resulted in a shortened developmental period for sixth-instar *S. exigua*, with a pooled mean effect size of −0.4334 (CI: −0.4744, −0.3923; Figure S6). Within the temperature range of 20–36 °C, the developmental duration of sixth-instar *S. exigua* significantly decreased as temperature increased (Figure 5A). The optimal condition for sixth-instar development was identified as 36 °C (Figure 5D).

Across all studies, higher temperatures led to a shortened developmental period for *S. exigua* eggs and the pooled mean effect size was calculated as −1.4807 (CI: −1.7341, −1.2274; Figure S7). The developmental time of *S. exigua* eggs increased initially and then decreased as the temperature rose within the range of 15.5–40 °C (Figure 6A). The results from both the random-effects and fixed-effects models showed Q (df = 18) = 4085.3176, $p < 0.0001$, suggesting that variations between studies had a substantial effect on the cumulative effect size, highlighting the need to include explanatory factors like humidity and photoperiod (Figure 6B). The ideal conditions for the development of *S. exigua* eggs were determined to be 36 °C, 65% relative humidity, and a 16:8 light/dark photoperiod. In all studies, elevated temperatures led to a reduction in the larval duration of *S. exigua*, with an overall mean effect size of −0.0997 (CI: −0.2942, −0.1012; Figure S8). Within the 22–28 °C temperature range, as the temperature rose, the larval duration of *S. exigua* was reduced (Figure 6A). The results from both the random-effects and fixed-effects models indicated Q (df = 26) = 9635.8421, $p < 0.0001$, suggesting considerable heterogeneity across studies, which affected the cumulative effect size. This highlights the necessity of incorporating explanatory variables (Figure 6C). The cumulative effect size was influenced by both photoperiod and relative humidity, with each factor impacting the larval period of *S. exigua* to different extents. The longest larval duration occurred at 28 °C, 60% relative humidity, and a 12:12 photoperiod.

In all studies, elevated temperatures resulted in a reduced pupal period for *S. exigua*, with a pooled mean effect size of −1.4421 (CI: −1.5845, −1.2998; Figure S9). Within the 15.5–38 °C temperature range, the pupal period duration significantly shortened with rising temperatures (Figure 6A). The results from both random-effects and fixed-effects models revealed Q (df = 51) = 1928.5778, $p < 0.0001$, suggesting that variations between studies impacted the cumulative effect size, emphasizing the need for explanatory variables (Figure 6D). The results indicate that relative humidity and photoperiod are key factors influencing the cumulative effect size, with different humidity levels exerting varying effects on the pupal duration of *S. exigua*. The shortest pupal developmental period was observed at 36 °C, with 70% relative humidity and a photoperiod of 14:10.

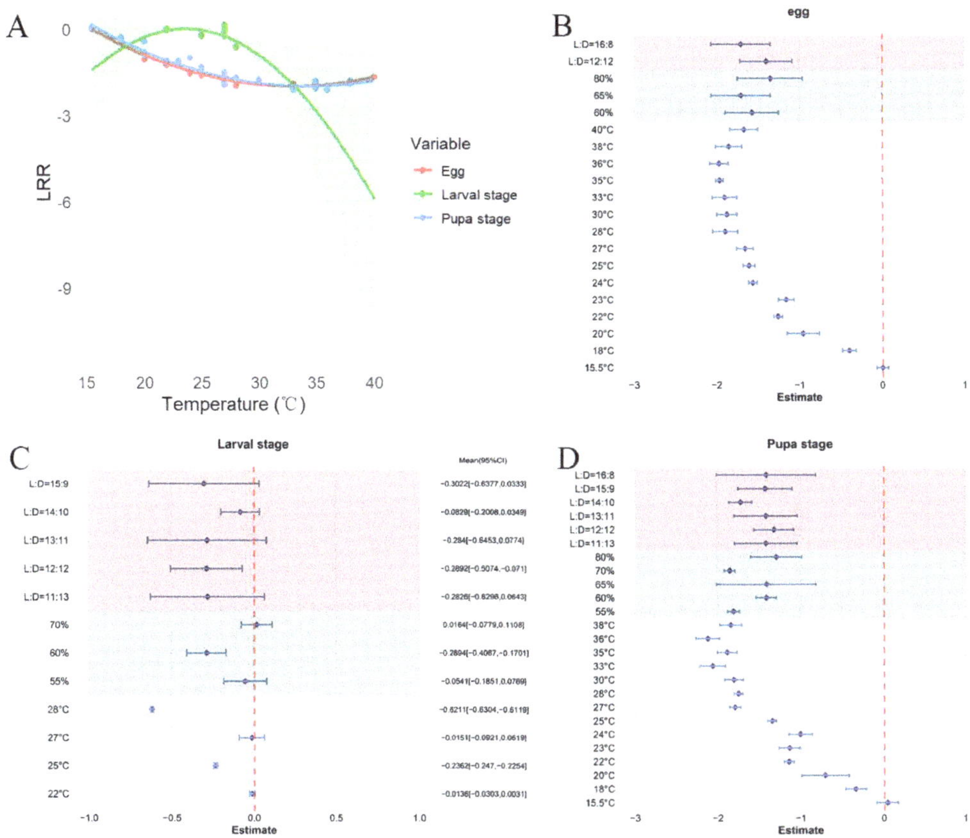

Figure 6. The effect of temperature on *S. exigua* egg, larval, and pupal periods. (**A**) shows the changes in developmental time for eggs, larvae, and pupae across different temperatures. (**B**–**D**) illustrate the response of eggs, larvae, and pupae to varying external environmental conditions.

3.4. The Influence of Temperature on the Ovipositional Behavior of Female Adults

Across all analyses, elevated temperatures were found to decrease the average egg-laying capacity of female *S. exigua*, with a pooled mean effect size of 0.2277 (CI: −0.0271, −0.4825; Figure S10). Across the temperature range of 15.5–40 °C, the egg-laying capacity of female *S. exigua* initially increased and then decreased as the temperature rose (Figure 7A). Results from both the random-effects and fixed-effects models showed Q (df = 15) = 113.0038, $p < 0.0001$, revealing significant heterogeneity between studies that affected the cumulative effect size, thereby necessitating the inclusion of explanatory variables (Figure 7B). The maximum egg-laying capacity was recorded at 24 °C, 70% relative humidity, and a photoperiod of 14:10.

Across all studies, higher temperatures were found to reduce the pre-oviposition period of female *S. exigua*, with a pooled mean effect size of −0.0966 (CI: −0.2508, −0.0576; Figure S11). Within the temperature span of 20–36 °C, the pre-oviposition period initially increased and then decreased as temperature rose (Figure 7A). The shortest pre-oviposition period for adult female *S. exigua* was observed at 36 °C (Figure 7C).

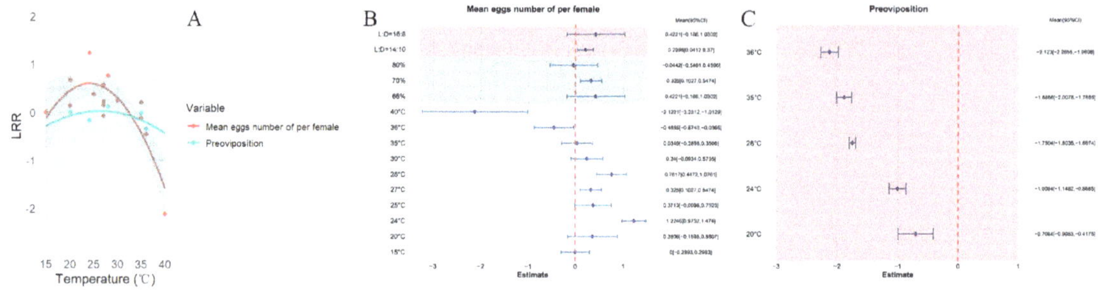

Figure 7. The effect of temperature on the ovipositional behavior and pre-oviposition period of female *S. exigua* adults. (**A**) shows how ovipositional behavior and the pre-oviposition period of female adults change with varying temperature. (**B**,**C**) illustrate the response of ovipositional behavior and the pre-oviposition period to different external environmental conditions.

3.5. The Effect of Temperature on Generation Time and Adult Lifespan

Across all the studies, higher temperatures were associated with a reduction in the adult lifespan of *S. exigua*, yielding a combined mean effect size of −0.6286 (CI: −0.7500, −0.5072; Figure S12). Throughout the temperature spectrum of 10–41 °C, the lifespan of adult *S. exigua* showed a significant reduction with rising temperatures (Figure 8A). Both the random-effects and fixed-effects models revealed substantial heterogeneity across studies (Q (df = 29) = 616.2937, $p < 0.0001$), which impacted the cumulative effect size and necessitated the inclusion of explanatory variables (Figure 8B). It is important to note that the photoperiod (QM = 17.4313, $p = 0.0002$) significantly influenced the cumulative effect size, showing differing effects on the lifespan of adult *S. exigua*. The shortest lifespan occurred at 36 °C, with 55% relative humidity and a 14:10 light-dark photoperiod.

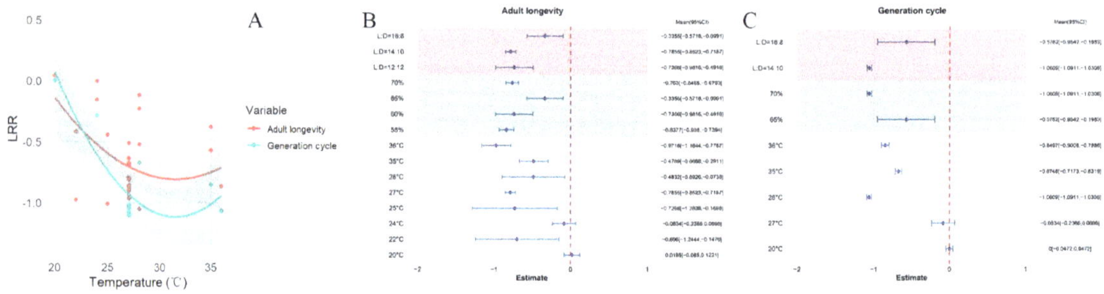

Figure 8. The effect of temperature fluctuations on the lifespan of *S. exigua* is presented as follows: (**A**) demonstrates the changes in the adult lifespan and total lifecycle duration with increasing temperature. (**B**) illustrates how the adult lifespan varies in response to changes in external environmental factors. (**C**) depicts the influence of environmental changes on the lifecycle duration of *S. exigua*.

Across the various studies, rising temperatures led to a shortened lifecycle of *S. exigua*, with a pooled mean effect size of −0.8175 (CI: −1.0058, −0.5793; Figure S13). Within the temperature span of 20 to 36 °C, the lifecycle of *S. exigua* consistently shortened as the temperature increased (Figure 8A). Results from both the random-effects and fixed-effects models indicated Q (df = 9) = 2513.7723, $p < 0.0001$, revealing significant heterogeneity among the studies, which affects the overall effect size and requires the inclusion of explanatory variables (Figure 8C). Notably, the photoperiod (QM = 6.2278, $p = 0.0126$) and relative humidity (QM = 6.2278, $p = 0.0126$) are identified as factors that affect the cumulative effect size, with varying degrees of influence on the lifecycle of *S. exigua*.

The shortest lifecycle was recorded at 36 °C, with 70% relative humidity and a 14:10 light-dark ratio.

3.6. Model Validation

We assessed the possibility of publication bias using both funnel and radar charts, and assessed the robustness of our results by calculating the fail-safe N. The findings from the funnel plot (Figure 9A; z = 3.6893, p = 0.0002), radar plot (Figure 9B), and fail-safe N (N = 27,385) all supported the reliability of our conclusions.

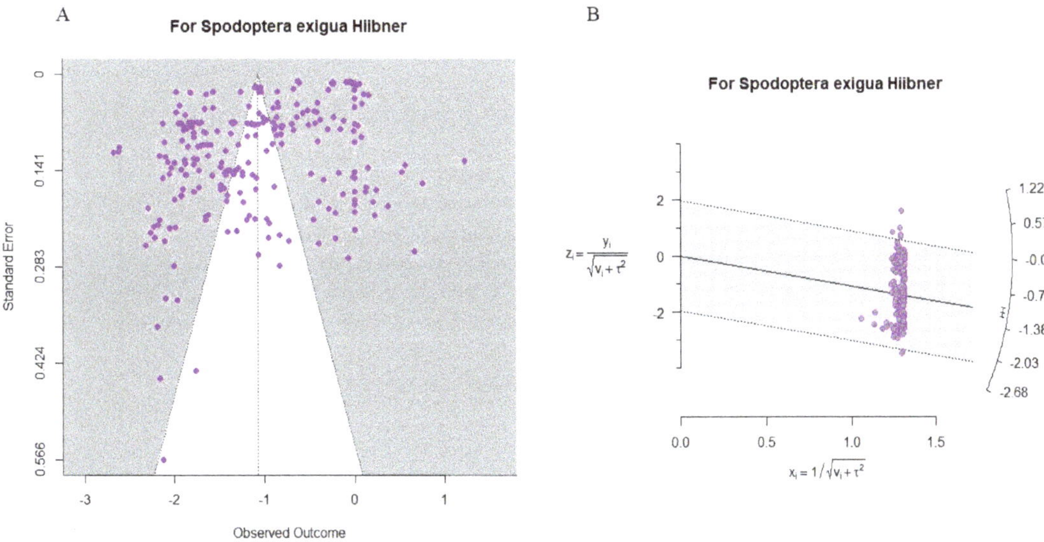

Figure 9. (**A**) Funnel plot; (**B**) radar plot.

4. Discussion

In the context of global warming, it is expected that the distribution of *S. exigua* will significantly expand. To assess its adaptability to higher temperatures, this study conducted a comprehensive meta-analysis of 33 relevant publications. Through a rigorous literature selection and data extraction process, we quantified the variation in *S. exigua*'s responses under different temperature conditions. The results show that the adaptability of *S. exigua* improves progressively with increasing temperatures, with a marked enhancement observed particularly between 15 °C and 40 °C [28,29]. At 33 °C, *S. exigua* reaches its peak adaptability, a finding that is consistent with previous studies and reinforces the reliability of the analytical model used [30,31]. Generally, a longer lifespan in insects tends to be associated with an extended reproductive period, offering more opportunities for egg laying. However, under the environmental conditions at 33 °C, the adult lifespan is reduced, leading to a shorter reproductive cycle, which in turn decreases egg production and results in lower fecundity.

Given the significant potential economic impact that *S. exigua* may have on agricultural production, monitoring its adaptability is crucial. Elevated temperatures generally expedite the developmental processes of insects, with growth rates strongly influenced by environmental conditions. Higher temperatures can shorten the developmental periods of various life stages—eggs, larvae, pupae, and adults [32,33]. Consequently, higher temperatures are anticipated to shorten the total lifecycle duration of *S. exigua*. Through meta-analysis, this study indicates that the species shows peak adaptability at temperatures around 33 °C.

Given the significance of early warning systems for tracking insect invasion routes and identifying appropriate environmental conditions, the results offer essential information to guide future monitoring and alert strategies.

As ectothermic organisms, insect populations are highly influenced by temperature, which plays a central role in their distribution patterns. With the accelerating effects of global warming, many insect species have experienced shifts in their geographical ranges, leading to a broader distribution [34–36]. Temperature is a crucial factor affecting the survival, development, and reproduction of *S. exigua*. Studies have shown that an optimal temperature range can significantly promote its growth and development, while temperature fluctuations can disrupt its growth cycle and reduce reproductive capacity. To investigate this connection in more detail, we analyzed the effect of temperature on different developmental stages of *S. exigua* and created the corresponding response curves. The results reveal that temperature affects each developmental stage differently, with distinct responses observed at each stage of growth.

The continuous rise in global temperatures is expected to significantly affect ecosystem function and structure, resulting in shifts in the distribution of biological habitats [37–39]. Climate change is expected to increase both the frequency and geographic spread of insect infestations, leading to greater agricultural damage. It is expected that the adaptability of *S. exigua* will enhance over time under future warming scenarios. The findings of this study offer important insights for policymakers to formulate more effective pest management strategies. These strategies will be essential in mitigating the potential economic losses that climate change may inflict on agricultural production in the future.

5. Conclusions

Spodoptera exigua is a significant agricultural pest that poses a serious threat to a wide range of crops. As such, developing an effective prediction model to monitor its presence in the field is essential. Conventional prediction approaches frequently use developmental parameters that depend on temperature and models of thermal biology to forecast changes in insect phenology. This study evaluated the adaptability of *S. exigua* to various temperature conditions by synthesizing data from 264 temperature-dependent experiments and performing a meta-analysis to predict its optimal temperature range for survival. Our results indicate that the adaptability of *S. exigua* improves as temperatures rise within the range of 15–40 °C, with optimal growth conditions occurring between 30 and 35 °C. Although the distribution of *S. exigua* has expanded over recent decades, current management strategies remain inadequate. It is essential to emphasize the significance of prompt chemical control. By predicting the optimal temperature range for *S. exigua* and preparing the appropriate chemical agents in advance, pest control efforts can be more effective, reducing its impact on agricultural production.

Supplementary Materials: The following supporting information can be downloaded at: https://www.mdpi.com/article/10.3390/insects16020155/s1, Figure S1: Comparison of effect sizes between models using random-effects and fixed-effects approaches; Figure S2: Comparison of effect sizes between models using random-effects and fixed-effects approaches; Figure S3: Comparison of effect sizes between models using random-effects and fixed-effects approaches; Figure S4: Comparison of effect sizes between models using random-effects and fixed-effects approaches; Figure S5: Comparison of effect sizes between models using random-effects and fixed-effects approaches; Figure S6: Comparison of effect sizes between models using random-effects and fixed-effects approaches; Figure S7: Comparison of effect sizes between models using random-effects and fixed-effects approaches; Figure S8: Comparison of effect sizes between models using random-effects and fixed-effects approaches; Figure S9: Comparison of effect sizes between models using random-effects and fixed-effects approaches; Figure S10: Comparison of effect sizes between models using random-effects and fixed-effects approaches; Figure S11: Comparison of effect sizes between models using random-effects and fixed-effects approaches; Figure S12: Comparison of effect sizes between models using random-effects and fixed-effects approaches; Figure S13: Comparison of effect sizes between models using random-effects and fixed-effects approaches.

Author Contributions: Conceptualization, Z.Z. and J.W.; methodology, H.Z. and X.D.; software, Z.L.; formal analysis, H.Z. and Z.L.; investigation, X.D. and Z.H.; data curation, Z.H. and J.W.; writing-original draft preparation, H.Z. and D.X.; writing-review and editing, Z.L. and Z.Z.; supervision, Z.Z. and D.X. All authors have read and agreed to the published version of the manuscript.

Funding: This work was funded by the Fundamental Research Funds of China West Normal University (20A007, 20E051, 21E040 and 22kA011).

Data Availability Statement: The data that support the findings can be accessed in a public repository at: https://doi.org/10.6084/m9.figshare.28057121.v2 (accessed on 19 December 2024).

Conflicts of Interest: The authors declare no conflicts of interest.

References

1. Lehmann, P.; Ammunét, T.; Barton, M.; Battisti, A.; Eigenbrode, S.D.; Jepsen, J.U.; Kalinkat, G.; Neuvonen, S.; Niemelä, P.; Terblanche, J.S.; et al. Complex responses of global insect pests to climate warming. *Front. Ecol. Environ.* **2020**, *18*, 141–150. [CrossRef]
2. Ramsfield, T.D.; Bentz, B.; Faccoli, M.; Jactel, H.; Brockerhoff, E.G. Forest health in a changing world: Effects of globalization and climate change on forest insect and pathogen impacts. *Forestry* **2016**, *89*, 245–252. [CrossRef]
3. Barford, E. Crop pests advancing with global warming. *Nature* **2013**. [CrossRef]
4. Youngblood, J.P.; Cease, A.; Talal, S.; Copa, F.; Medina, H.E.; Rojas, J.E.; Trumper, E.V.; Angilletta, M.J.; Harrison, J.F. Climate change expected to improve digestive rate and trigger range expansion in outbreaking locusts. *Ecol. Monogr.* **2022**, *93*, e1550. [CrossRef]
5. Huang, Y.; Dong, Y.; Huang, W.; Guo, J.; Hao, Z.; Zhao, M.; Hu, B.; Cheng, X.; Wang, M. Predicting the Global Potential Suitable Distribution of Fall Armyworm and Its Host Plants Based on Machine Learning Models. *Remote Sens.* **2024**, *16*, 2060. [CrossRef]
6. Sandra, S.; Monika, Z.; Pajač, Ž.I.; Vinko, L.; Darija, L. The Impact of Climate Change on Agricultural Insect Pests. *Insects* **2021**, *12*, 440. [CrossRef] [PubMed]
7. Chowdary, L.R.; Kumar, G.V.S.; Bharathi, S.; Sarada, Q.; Nagaraju, Y.; Chandrashekara, K.M.; Harish, G.N. Off-season survival and life history of beet armyworm, *Spodoptera exigua* (Hubner) on various host plants. *Sci. Rep.* **2024**, *14*, 13721.
8. Zheng, X.; Huang, Q.; Cao, W.; Lu, W.; Wang, G.; Yu, S.; Yang, Z.; Wang, X. Modeling climate change impacts on overwintering of *Spodoptera exigua* Hübner in regions of China. *Chil. J. Agr. Res.* **2015**, *75*, 328–333. [CrossRef]
9. Breeschoten, T.; Ros, V.I.D.; Schranz, M.E.; Simon, S. An influential meal: Host plant dependent transcriptional variation in the beet armyworm, *Spodoptera exigua* (Lepidoptera: Noctuidae). *BMC Genom.* **2019**, *20*, 845. [CrossRef]
10. Zhang, B.; Liu, H.; Helen, H.; Wang, J. Effect of host plants on development, fecundity and enzyme activity of *Spodoptera exigua* (Hübner) (Lepidoptera: Noctuidae). *Agri. Sci. China* **2011**, *10*, 1232–1240. [CrossRef]
11. Mehrkhou, F.; Talebi, A.A.; Moharramipour, S.; Naveh, V.H.; Farahani, S. Development and fecundity of *Spodoptera exigua* (Hübner) (Lepidoptera: Noctuidae) on different soybean cultivars. *Arch. Phytopathol. Plant Prot.* **2012**, *45*, 90–98. [CrossRef]
12. Dai, H.; Zhang, G.; Zhang, W. Temperature dependent development parameters and population life table of beet armyworm, *Spodoptera exigua* (Hübner) (Lepidoptera: Noctuidae). *Arthropods* **2017**, *6*, 117–125.
13. Ma, H.T.; Zhou, L.H.; Tan, H.; Xiu, X.Z.; Wang, J.Y.; Wang, X.Y. Population dynamics and seasonal migration patterns of *Spodoptera exigua* in northern China based on 11 years of monitoring data. *PeerJ* **2024**, *12*, e17223. [CrossRef] [PubMed]
14. Zhou, L.H.; Wang, X.Y.; Xu, G.Q.; Lei, J.J. Mitochondrial DNA phylogeography of *Spodoptera exigua* across a broad geographic area in China. *J. Appl. Entomol.* **2017**, *141*, 527–539. [CrossRef]
15. Jia, H.; Wang, T.; Li, X.; Zhao, S.; Guo, J.; Liu, A.; Liu, Y.; Wu, K. Pollen molecular identification from a long-distance migratory insect, *Spodoptera exigua*, as evidenced for its regional pollination in Eastern Asia. *Int. J. Mol. Sci.* **2023**, *24*, 7588. [CrossRef]
16. Lemoine, N.P.; Shantz, A.A. Increased temperature causes protein limitation by reducing the efficiency of nitrogen digestion in the ectothermic herbivore *Spodoptera exigua*. *Physiol. Entomol.* **2016**, *41*, 143–151. [CrossRef]
17. Zheng, X.L.; Wang, P.; Cheng, W.J.; Lu, W.; Xian, Z.H.; Lei, C.L.; Wang, X.P. Effects of the larval host plant on the supercooling capacity and physiological characteristics of beet armyworm pupae, *Spodoptera exigua* (Lepidoptera: Noctuidae). *J. Plant Dis. Protect* **2014**, *121*, 202–210. [CrossRef]

18. Fu, C.; Liu, Z.; Xu, D.; Peng, Y.; Liu, B.; Zhuo, Z. Effects of Global Climate Warming on the Biological Characteristics of *Spodoptera frugiperda* (J.E. Smith) (Lepidoptera: Noctuidae). *Insects* **2024**, *15*, 689. [CrossRef] [PubMed]
19. Hardwicke, T.E.; Serghiou, S.; Janiaud, P.; Danchev, V.; Crüwell, S.; Goodman, S.N.; Ioannidis, J.P. Calibrating the scientific ecosystem through meta-research. *Annu. Rev. Stat. Appl.* **2020**, *7*, 11–37. [CrossRef]
20. Viechtbauer, W. Conducting Meta-Analyses in R with the metafor Package. *J. Stat. Softw.* **2010**, *36*, 1–48. [CrossRef]
21. Bo, C.; Andrea, B. Quantifying heterogeneity in individual participant data meta-analysis with binary outcomes. *Syst. Rev.* **2017**, *6*, 243.
22. Thorlund, K.; Imberger, G.; Johnston, B.C.; Walsh, M.; Awad, T.; Thabane, L.; Gluud, C.; Devereaux, P.J.; Wetterslev, J. Evolution of heterogeneity (I2) estimates and their 95% confidence intervals in large meta-analyses. *PLoS ONE* **2017**, *7*, e39471. [CrossRef] [PubMed]
23. McKenzie, J.E.; Veroniki, A.A. A brief note on the random-effects meta-analysis model and its relationship to other models. *J. Clin. Epidemiol.* **2024**, *174*, 111492. [CrossRef] [PubMed]
24. Plonsky, L.; Oswald, F.L. How to do a meta-analysis. In *Research Methods in Second. Language Acquisition: A Practical Guide*; Blackwell Publishing Ltd.: Oxford, UK, 2011; pp. 275–295.
25. Hedges, L.V.; Olkin, I. *Statistical Methods for Meta-Analysis*; Academic Press: Cambridge, MA, USA, 2014; pp. 75–78.
26. Egger, M.; Smith, G.D.; Schneider, M.; Minder, C. Bias in meta-analysis detected by a simple, graphical test. *BMJ-Brit Med. J.* **1997**, *315*, 629–634. [CrossRef] [PubMed]
27. Philip, S.; Louise, M. How to read a funnel plot in a meta-analysis. *BMJ (Clin. Res. Ed.)* **2015**, *351*, h4718.
28. Lee, S.D.; Ahn, S.B.; Choi, K.M.; Cho, W.S. Effect of temperature on the development of beet armyworm, *Spodoptera exigua* Hubner (Lepidoptera: Noctuidae). *Res. Rep. Rural. Dev. Admin.* **1991**, *33*, 1013–9389.
29. Jiang, X.; Luo, L.; Li, K.; Cao, Y.; Hu, Y.; Liu, Y. Influence of temperature on flight capacity of the beet armyworm, *Spodoptera exigua*. *Acta Entomol. Sin.* **2002**, *45*, 275–278.
30. Karimi-Malati, A.; Fathipour, Y.; Talebi, A.A.; Bazoubandi, M. Life table parameters and survivorship of *Spodoptera exigua* (Lepidoptera: Noctuidae) at constant temperatures. *Environ. Entomol.* **2014**, *43*, 795–803. [CrossRef] [PubMed]
31. Karimi-Malati, A.; Fathipour, Y.; Talebi, A.A. Development response of *Spodoptera exigua* to eight constant temperatures: Linear and nonlinear modeling. *J. Asia-Pac. Entomol.* **2014**, *17*, 349–354. [CrossRef]
32. Liu, Z.; Liu, B.; Yu, H.; Zhang, H.; He, Z.; Zhuo, Z. The Effects of Global Climate Warming on the Developmental Parameters of *Helicoverpa armigera* (Hübner, 1808) (Lepidoptera: Noctuidae). *Insects* **2024**, *15*, 888. [CrossRef]
33. Ouyang, F.; Hui, C.; Men, X.; Zhang, Y.; Fan, L.; Shi, P.; Zhao, Z.; Ge, F. Early eclosion of overwintering cotton bollworm moths from warming temperatures accentuates yield loss in wheat. *Agric. Ecosyst. Environ.* **2016**, *217*, 89–98. [CrossRef]
34. Ulrichs, C.; Hopper, K.R. Predicting insect distributions from climate and habitat data. *Biocontrol* **2008**, *53*, 881–894. [CrossRef]
35. Jin, Z.; Zhao, H.; Xian, X.; Li, M.; Qi, Y.; Guo, J.; Yang, N.; Lü, Z.; Liu, W. Early warning and management of invasive crop pests under global warming: Estimating the global geographical distribution patterns and ecological niche overlap of three Diabrotica beetles. *Environ. Sci. Pollut. R.* **2024**, *31*, 13575–13590. [CrossRef] [PubMed]
36. Zeng, J.; Liu, Y.; Zhang, H.; Liu, J.; Jiang, Y.; Wyckhuys, K.A.G.; Wu, K. Global warming modifies long-distance migration of an agricultural insect pest. *J. Pest. Sci.* **2020**, *93*, 569–581. [CrossRef]
37. Botkin, D.B.; Saxe, H.; Araújo, M.B.; Betts, R.; Bradshaw, R.H.; Cedhagen, T.; Chesson, P.; Dawson, T.P.; Etterson, J.R.; Faith, D.P. Forecasting the effects of global warming on biodiversity. *Bioscience* **2007**, *57*, 227–236. [CrossRef]
38. Gao, H.; Wei, X.; Peng, Y.; Zhuo, Z. Predicting the Impact of Climate Change on the Future Distribution of Paederus fuscipes Curtis, 1826, in China Based on the MaxEnt Model. *Insects* **2024**, *15*, 437. [CrossRef] [PubMed]
39. Deng, X.; Xu, D.; Liao, W.; Wang, R.; Zhuo, Z. Predicting the distributions of *Scleroderma guani* (Hymenoptera: Bethylidae) under climate change in China. *Ecol. Evol.* **2022**, *12*, e9410. [CrossRef] [PubMed]

insects

MDPI

Article

Influence of Temperature, Humidity, and Photophase on the Developmental Stages of *Spodoptera litura* (Lepidoptera: Noctuidae) and Prediction of Its Population Dynamics

Chun Fu [1,†], Zhiqian Liu [2,†], Danping Xu [2], Tingjiang Gan [3], Xinqi Deng [2], Honghua Zhang [2] and Zhihang Zhuo [2,*]

1 Key Laboratory of Sichuan Province for Bamboo Pests Control and Resource Development, Leshan Normal University, Leshan 614000, China; fuchun421@aliyun.com
2 College of Life Science, China West Normal University, Nanchong 637002, China; qnhtvxhp319123@foxmail.com (Z.L.); xudanping@cwnu.edu.cn (D.X.); deng.xinqi@foxmail.com (X.D.); honghua_zhang@foxmail.com (H.Z.)
3 Engineering Research Centre of Chuanxibei Rural Human Settlement (RHS) Construction, Mianyang Teachers' College, Mianyang 621016, China; gantinjiangky@mtc.edu.cn
* Correspondence: zhuozhihang@foxmail.com; Tel.: +86-131-1197-3927
† These authors contributed equally to this work.

Simple Summary: *Spodoptera litura* (Fabricius, 1775) is a significant economic pest that has successfully invaded Africa and Asia in recent years. This study systematically evaluated the life history traits of *S. litura* under different temperature, photoperiod, and humidity conditions. The results showed that at 30–35 °C, the physiological activity of *S. litura* peaked, with a significant reduction in the duration of developmental stages, increased female oviposition, shortened pupal and adult lifespans, and an accelerated generational cycle. These findings provide critical insights into predicting population dynamics and offer valuable guidance for developing effective management strategies.

Abstract: *Spodoptera litura* (Fabricius, 1775) is a major agricultural pest that primarily targets vegetables, cash crops, peanuts, and sugarcane. It causes damage to leaves, flower buds, and fruits, leading to significant reductions in crop yields. Global climate change may profoundly affect the population dynamics and biological traits of this pest. This research employs a meta-analysis to systematically investigate the impact of temperature variation on the developmental parameters of *S. litura*. A detailed review of 17 relevant studies reveals that within an optimal temperature range (30 °C to 35 °C), higher temperatures expedite the developmental processes of *S. litura*, shorten its life cycle, and enhance the reproductive potential of female adults. In contrast, temperatures exceeding 35 °C slow down its development, increase mortality rates, and markedly reduce the egg-laying capacity of females, highlighting the adverse effects of heat stress on growth and reproduction. Furthermore, different life stages of *S. litura* exhibit varying degrees of temperature sensitivity, with the larval stage being particularly vulnerable to high temperatures, while extreme heat significantly suppresses adult survival. These meta-analysis findings shed light on the biological responses of *S. litura* to climate change and provide a scientific basis for developing future pest management strategies. As global temperatures rise, moderate warming may facilitate the spread of *S. litura* populations, exacerbating their threat to crop production, whereas extreme heat conditions could constrain their growth. Consequently, pest control strategies must be more region-specific and aligned with local climatic trends.

Keywords: *Spodoptera litura*; climate change; agricultural pests; invasive insects; meta-analysis

 https://doi.org/10.3390/insects16040355

1. Introduction

Climate change has become one of the most challenging environmental issues facing modern agricultural ecosystems, profoundly influencing the occurrence and development of crop pests and diseases [1,2]. Changes in climate factors such as temperature, precipitation, and atmospheric CO_2 concentration have not only altered the dynamics of agricultural ecosystems but also had far-reaching impacts on pest growth and development, reproductive capacity, migration behavior, and biodiversity [3–5]. In this context, *S. litura*, as a major pest in global agricultural production, may experience significant changes in its biological characteristics due to the effects of climate change, potentially posing a threat to agricultural production [6,7]. Investigating the impact of climate change on the biological characteristics of *S. litura* is crucial for developing effective pest control strategies.

Spodoptera litura (Fabricius, 1775) is widely distributed across agricultural regions globally, particularly in tropical and subtropical areas, where its larvae primarily feed on crops, especially maize, soybeans, and cotton [8]. As a polyphagous pest active both day and night, *S. litura* can adapt to various climatic conditions, and its growth, development, reproductive capacity, and migration behaviors are highly sensitive to climate change [9]. In recent years, with the intensification of global climate change, the distribution and frequency of *S. litura* occurrences have expanded, posing an increasing challenge to agricultural production [10]. Climate change impacts *S. litura*'s development rate, reproductive potential, survival rate, and population dynamics, directly or indirectly altering its biological characteristics [11,12]. Therefore, systematically assessing the combined effects of climate factors, particularly changes in temperature and humidity, on *S. litura* will provide scientific evidence for early pest warning systems and effective pest control strategies.

Existing studies have shown that temperature is a key factor influencing the biological characteristics of *S. litura* [13,14]. Higher temperatures typically accelerate its life cycle, shorten the duration of each developmental stage, increase generation turnover, and lead to population outbreaks [15]. However, excessively high temperatures may exceed the species' adaptive capacity, resulting in increased mortality or incomplete development [16]. Additionally, humidity and photoperiod may significantly impact the survival and distribution of *S. litura* [17]. Droughts or extreme precipitation events not only affect habitat suitability but may also alter the availability of food resources [18]. Therefore, comprehensively assessing the impact of these climate factors on *S. litura* will help provide more targeted strategies for pest management.

Although previous studies have explored the effects of climate change on the biological characteristics of *S. litura*, there are some discrepancies and uncertainties due to differences in research methods, climate scenarios, and regional contexts. Meta-analysis, as an effective statistical approach, can integrate the results of different studies, quantify the overall impact of climate change on the biological characteristics of *S. litura*, and reveal the magnitude and direction of the effects of various climate factors [19]. This study uses meta-analysis to address the following questions: How does climate change affect the development rate, reproductive potential, and survival rate of *S. litura*? The findings will not only contribute to a deeper understanding of the ecological responses of *S. litura* under climate change but also provide data support and theoretical foundations for future pest prediction models and control strategies.

2. Materials and Methods

2.1. Literature Search

In this study, we employed a systematic literature review approach, gathering relevant research from various databases [20]. The literature search was conducted between August and October 2024, primarily focusing on Web of Science, PubMed, Scopus, and CNKI.

Additionally, the reference lists of related review articles were manually examined by us to identify studies not indexed in these databases. The search terms included "*Spodoptera litura*", "climate change" (including "global warming" or "climate change"), "temperature", "humidity" (or "precipitation"), and "biological traits" (such as "biological traits", "life history traits", "development", or "reproduction"). Boolean operators (AND, OR) were used to ensure comprehensive coverage of the relevant literature [21].

The selection process was conducted in two steps. In the first step, titles and abstracts were screened to exclude studies unrelated to climate change or *S. litura*. In the second step, a full-text review was performed to exclude studies that did not meet the inclusion criteria. The inclusion criteria required the study to explicitly investigate the effects of climate factors, such as temperature and humidity, on biological traits of *S. litura* (e.g., developmental rate, reproductive capacity, survival rate). Studies were also required to include a control group and provide clear documentation of experimental conditions (e.g., temperature range, humidity levels). Additionally, the study had to provide data suitable for meta-analysis (e.g., means, standard deviations, sample sizes) or data that could be extracted from figures. Exclusion criteria included studies lacking a control group, those based solely on theoretical models, studies unrelated to *S. litura* biological traits or climate factors, and studies that did not provide sufficient statistical data for meta-analysis.

2.2. Data Extraction

Data were gathered from studies that met the inclusion criteria. The key variables included temperature (experimental treatment temperatures in °C), relative humidity, and biological traits (such as development rate, generation time, and reproductive metrics like egg-laying ability). For studies lacking direct statistical details, such as standard deviations or sample sizes, the required data were retrieved from figures using graphical extraction tools like WebPlotDigitizer (v5). In cases of multivariate experiments or studies with multiple treatment groups, each treatment was treated as an independent effect size for analysis.

2.3. Statistical Analysis

The analysis was conducted using the "rma.mv" function from the "metafor" package in R version 4.3 [22]. A random-effects model was applied to calculate the relative risk (RR+) and estimate variance (I^2) between cases using restricted maximum likelihood (REML) [23–25]. Explanatory variables were added to the model based on the I^2 value. The overall mean effect size of temperature across all treatment groups was determined using a random-effects model [26]. Statistical tests, including mean effect size, 95% confidence intervals (CI), Qt, and I^2, were performed. "Qt" is used to represent the total heterogeneity statistic (the overall value of Cochran's Q) after combining all included studies, which helps to assess whether there are significant differences between the study results.

Meta-analyses were conducted to assess the impact of temperature on various biological traits of *S. litura* across different developmental stages. Both random-effects and fixed-effects models were used, with the random-effects model preferred due to significant heterogeneity in experimental locations and methods [27].

In this analysis, temperature and relative humidity were treated as independent variables, while the biological traits of *S. litura* were the dependent variables. Effect sizes were calculated using the Log Response Ratio (LRR) [28], with statistical significance assessed through 95% confidence intervals. Subgroup analyses were performed to examine the influence of temperature and relative humidity on *S. litura*'s biological traits.

Heterogeneity was assessed using Q statistics and the I^2 index [29]. The I^2 value indicates the extent of variability, with higher values signifying greater heterogeneity.

The heterogeneity statistic was calculated by testing the weighted sum of squares based on a k-1 distribution. If the 95% confidence interval of the effect size includes 0, there is no significant difference between the experimental and control groups ($p > 0.05$). If the confidence interval is entirely above 0, the experimental group shows a significantly larger effect size ($p < 0.05$). Conversely, if the confidence interval is entirely below 0, the experimental group shows a smaller effect size ($p < 0.05$).

Explanatory variables were included based on the significance of the cumulative effect size relative to zero and the *p*-value of Qt. Key variables considered included the effects of humidity and photoperiod on the cumulative effect size. Temperature was treated as a continuous variable to assess its impact on the mean effect size. In the meta-analysis, heterogeneity was divided into between-group variance (explained by categorical factors) and within-group residual variance, with statistical significance assessed using a k-1 test. Publication bias was evaluated using funnel plots and radar charts [30,31]. If bias was detected, the "trim and fill" method was used to correct it [32].

3. Results

3.1. Literature Search and Screening Results

A comprehensive search identified 751 studies that could potentially be relevant. In the initial screening, 508 studies not related to the research topic were removed. After a thorough full-text review of the remaining 243 studies, 17 studies that fulfilled the inclusion criteria were selected. These studies investigated the impact of various climate factors, such as temperature and humidity, on the biological traits of *S. litura*. A total of 857 datasets were extracted from these 17 studies, covering the following parameters: 1st instar ($n = 43$), 2nd instar ($n = 43$), 3ird instar ($n = 43$), 4th instar ($n = 43$), 5th instar ($n = 43$), 6th instar ($n = 43$), adult longevity (female), $n = 51$), adult longevity (male), $n = 47$), adult stage ($n = 26$), egg ($n = 105$), egg-to-adult ($n = 16$), generation ($n = 28$), oviposition ($n = 8$), pre-oviposition ($n = 35$), pre-pupa ($n = 15$), pupa ($n = 116$), fertility ($n = 31$), and larval stage ($n = 121$). In addition, the overall impact of temperature changes on *S. litura* was analyzed (Table 1).

Table 1. Data volume of *S. litura* after screening. (TR represents the temperature range, CT represents the control group temperature, and *n* represents the sample size).

TR	CT	*n*	Variable
17–34 °C	17 °C	43	1st instar
17–34 °C	17 °C	43	2nd instar
17–34 °C	17 °C	43	3ird instar
17–34 °C	17 °C	43	4rth instar
17–34 °C	17 °C	43	5th instar
17–34 °C	17 °C	43	6th instar
15–35 °C	15 °C	51	Adult longevity (Female)
15–35 °C	15 °C	47	Adult longevity (male)
17–35 °C	17 °C	26	Adult stage
15–38 °C	15 °C	105	Egg
17–33 °C	17 °C	16	Egg to adult
17–35 °C	17 °C	28	Generation
17–33 °C	17 °C	8	Oviposition
17–34 °C	17 °C	35	Pre-oviposition
17–35 °C	17 °C	15	Pre-pupa
15–38 °C	15 °C	116	Pupa
17–35 °C	17 °C	31	Fertility
15–38 °C	15 °C	121	Larval stage
15–38 °C	15 °C	857	*Spodoptera litura*

3.2. Overall Effect of Temperature on the Biological Traits of *S. litura*

After synthesizing the data from various studies and performing a meta-analysis, the findings suggest that increasing temperature increases the adaptability of *S. litura*. The overall mean effect size is −0.8077 (CI: −0.8509; −0.7645; Figure 1, Table 2, Figure S1). As temperature rises, the duration of the oviposition period decreases, the oviposition rate of female adults increases, egg hatching time shortens, the development period of 1st to 5th instar larvae is reduced, and adult longevity increases. However, there is no significant effect on the development period of the 6th instar larvae or the pre-oviposition period (Figure 2). By considering temperature as a continuous factor, the overall impact of temperature variations on *S. litura* across different temperature ranges was observed. With rising temperature, all physiological parameters of *S. litura* significantly increased (Figure 3A). As the temperature rises, the physiological activity of *S. litura* peaks at 35 °C. However, when temperatures exceeded 35 °C, there was a noticeable decline in all physiological indicators (Figure 3B).

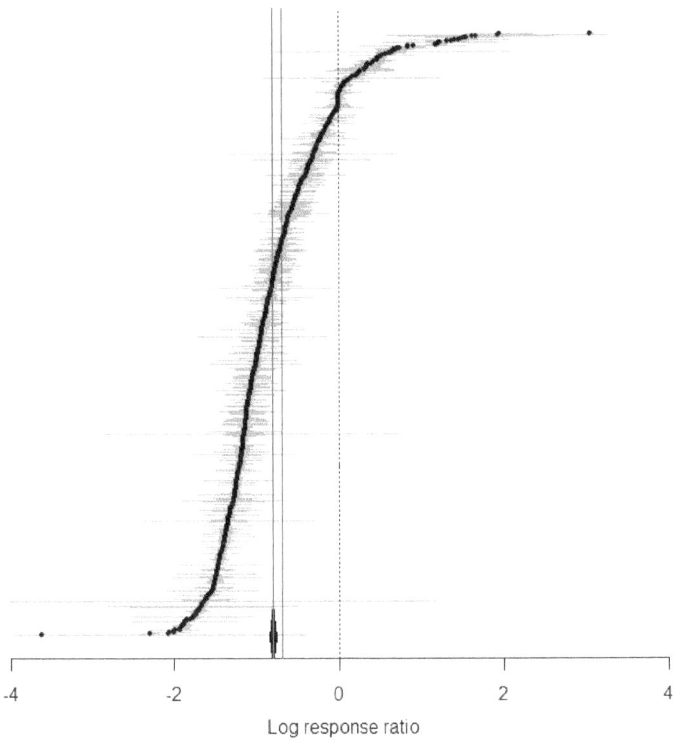

Figure 1. The effect of temperature changes on *S. litura* is presented. (Red line indicates the result derived from the random-effects model, showing a total effect size of −0.8077 with a 95%CI ranging from −0.8509 to −0.7645. In comparison, Black solid line represents the fixed-effects model result, with a total effect size of −0.6888 and a 95%CI of −0.6898 to −0.6878. The blue solid line shows the size of the cumulative effect).

Table 2. Results of subgroup analysis calculated using the random-effects model.

Variable	Estimate	SE	Z	*p*	CI.lb	CI.ub	Loglik	AIC	BIC
1st instar	−1.1675	0.0594	−19.6663	<0.0001	−1.2839	−1.0512	−19.9416	43.8832	47.3585
2nd instar	−0.8725	0.0619	−14.0993	<0.0001	−0.9938	−0.7512	−21.6789	47.3578	50.8331

Table 2. *Cont.*

Variable	Estimate	SE	Z	*p*	CI.lb	CI.ub	Loglik	AIC	BIC
3ird instar	−0.8763	0.0699	−12.5428	<0.0001	−1.0132	−0.7393	−26.8192	57.6384	61.1137
4th instar	−0.72	0.1088	−6.6187	<0.0001	−0.9332	−0.5068	−45.3916	94.7831	98.2584
5th instar	−0.8143	0.0795	−10.2486	<0.0001	−0.9701	−0.6586	−32.1914	68.3829	71.8582
6th instar	0.0597	0.0850	0.7018	0.4828	−0.1070	0.2263	−35.2275	74.4549	77.9303
Adult longevity (Female)	−0.9734	0.0476	−20.4557	<0.0001	−1.0667	−0.8801	−17.2454	38.4908	42.3148
Adult longevity (male)	−1.1394	0.0558	−20.441	<0.0001	−1.2488	−1.03	−20.7576	45.5151	49.1724
Adult stage	−1.0392	0.1132	−9.1778	<0.0001	−1.2611	−0.8173	−21.6633	47.3266	49.7644
Egg	−1.2022	0.054	−22.2472	<0.0001	−1.3081	−1.0963	−85.7709	175.5419	180.8306
Egg to adult	−0.3387	0.0875	−3.8707	0.0001	−0.5102	−0.1672	−5.4890	14.9779	16.3940
Generation	−0.9917	0.1043	−9.5097	<0.0001	−1.1961	−0.7873	−22.2479	48.4958	51.0874
Oviposition	−0.9705	0.2411	−4.0252	<0.0001	−1.4434	−0.4979	−7.2267	18.4534	18.3452
Pre−oviposition	−0.092	0.083	−1.1081	0.2678	−0.2546	0.0707	−24.5303	53.0607	56.1134
Pre−pupa	−1.0001	0.1071	−9.3409	<0.0001	−1.2099	−0.7902	−7.5383	19.0766	20.3547
Pupa	−0.9870	0.0432	−22.8742	<0.0001	−1.0716	−0.9025	−74.7748	153.5496	159.0394
Fertility	0.4685	0.1958	2.3926	0.0167	0.0847	0.8523	−46.5657	97.1315	99.9339
Larval stage	−0.6795	0.0426	−15.9436	<0.0001	−0.763	−0.596	−79.0335	162.0671	167.642
Spodoptera litura	−0.8077	0.022	−36.6432	<0.0001	−0.8509	−0.7645	−849.4961	1702.9922	1712.4967

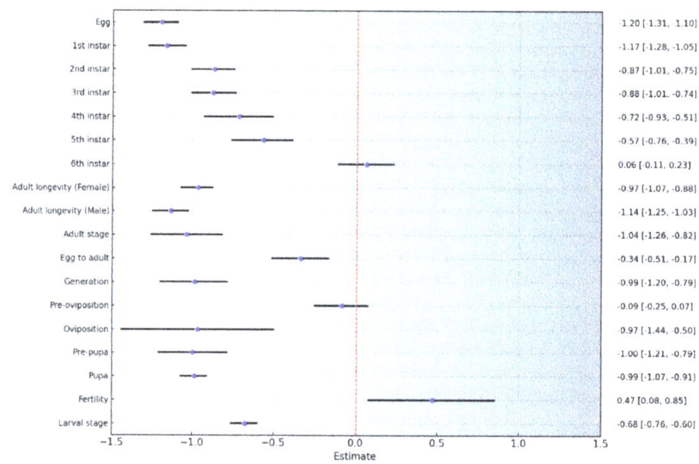

Figure 2. The effect of temperature changes on various physiological parameters of *S. litura*. (Blue dots represent the total effect size, and the black solid lines represent the 95% upper and lower confidence intervals).

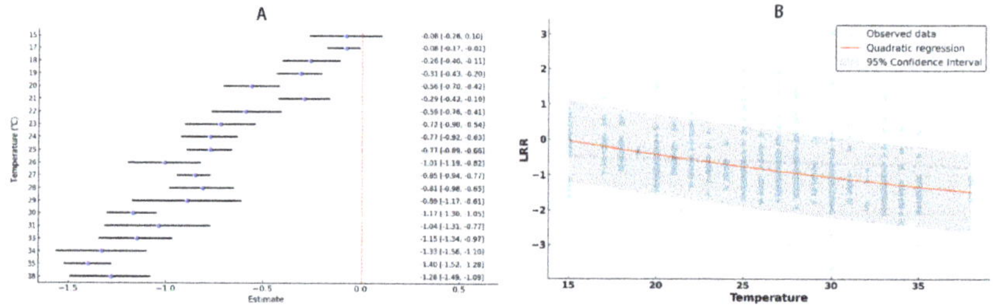

Figure 3. The impact of temperature changes on the developmental rate of *S. litura* is illustrated. Panel (**A**) shows how the physiological characteristics of *S. litura* vary with increasing temperature. The magnitude of the mean value reflects the impact of temperature: smaller mean values indicate a faster developmental rate, while larger mean values suggest a slower rate. (**B**) presents the temperature range curve that identifies the optimal growth conditions for *S. litura*.

3.3. The Impact of Temperature on Development Period

Temperature has a significant impact on the hatching rate and developmental speed of *S. litura* eggs (Figure 4A), with an overall average effect size of −1.2022 (CI: −1.3081; −1.0963; Figure S2, Table 2). The temperature–response curve analysis revealed that 34 °C is the optimal temperature for egg development (Figure 4B). At this temperature, the hatching rate is highest, and the developmental speed is fastest, indicating that 34 °C provides the ideal conditions for egg growth. Both higher and lower temperatures result in a marked decline in hatching rate and developmental speed. Additionally, the optimal humidity for the egg stage was found to be 60%, with this relatively low humidity being sufficient to support normal egg development. Regarding photoperiod, the optimal light cycle for the egg stage was determined to be 12 h light:12 h dark, suggesting that *Spodoptera litura* eggs develop most effectively under a 12:12 light–dark cycle.

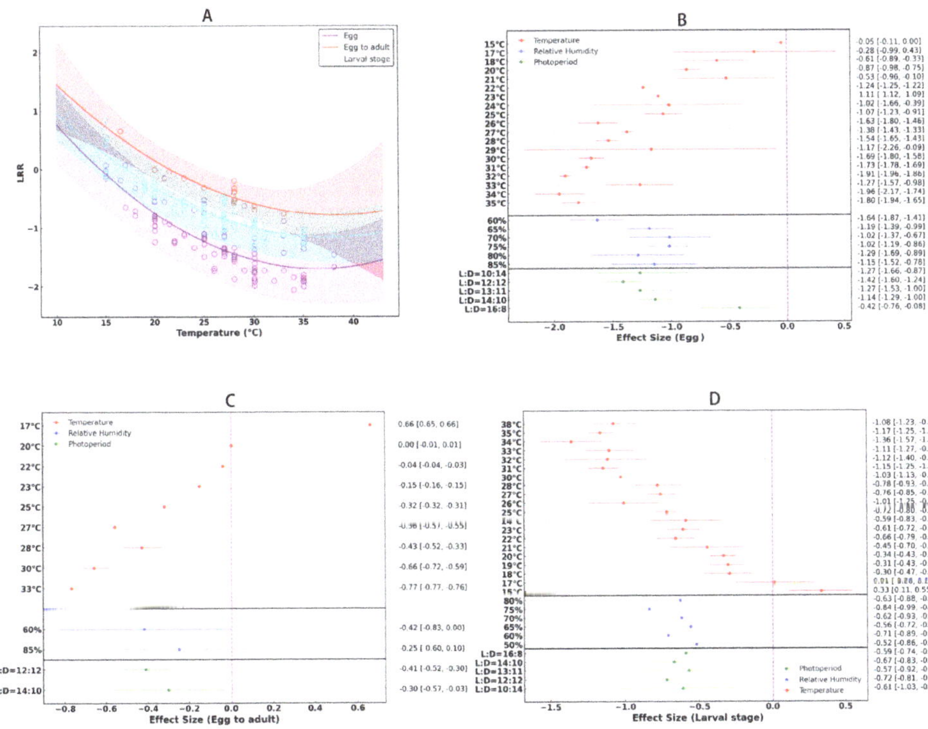

Figure 4. The effects of temperature on the developmental duration of *S. litura* eggs, egg-to-adult development, and larval stage are shown. Panel (**A**) illustrates the trend in developmental time as temperature increases. Panels (**B**–**D**) highlight the optimal environmental conditions for the development of eggs, egg-to-adult stages, and larvae, respectively.

For the egg to adult developmental stage, temperature has a significant impact on the hatching rate and developmental speed of *S. litura* egg to adult (Figure 4A), with an overall average effect size of −0.3387 (CI: −0.5102; −0.1672; Figure S3, Table 2). the optimal temperature slightly decreases to 33 °C (Figure 4C). At this temperature, development from egg to adult proceeds most smoothly, indicating that the temperature requirement for this stage is slightly lower than that of the egg stage. When temperatures deviate from this value, development is significantly inhibited, leading to lower hatching and survival rates. Humidity for this stage was also found to be 60%, similar to the egg stage, which effectively supports normal developmental processes. The optimal photoperiod for the egg to adult stage was also 12:12, consistent with the egg stage, highlighting the importance of the 12 h light and 12 h dark cycle for proper development during this phase.

Temperature continues to play a crucial role during the larval stage of *S. litura* (Figure 4A), with an overall average effect size of −0.6795 (CI: −0.7630; −0.5960; Figure S4, Table 2). The optimal temperature for larval development is 34 °C, providing the best conditions for growth and survival (Figure 4D). Similar to the egg stage, any deviation from 34 °C leads to a significant reduction in developmental speed and survival rates. In terms of humidity, the optimal level for the larval stage was notably higher than in the other stages, reaching 75%, which facilitates better growth and survival. This higher humidity level is critical for supporting healthy larval development. Like the egg and egg to adult stages, the larval stage also thrives under a 12:12 light cycle, indicating a strong dependence on a consistent photoperiod for proper growth and development.

3.4. Temperature Effects on Developmental Duration at Different Larval Stages

In the first instar of *Spodoptera litura*, temperature significantly influenced development, as shown by the temperature response curve analysis (Figure 5A), with an overall average effect size of −1.1675 (CI: −1.2839; −1.0512; Figure S5, Table 2). The optimal temperature for development was 33 °C (Figure 5B). At this temperature, the first instar larvae exhibited the fastest development and best growth. Temperatures that were either too high or too low resulted in delayed development and decreased survival rates, making 33 °C the ideal temperature for this stage. Additionally, a humidity of 76% provided an optimal environment for development, and a 14:10 light:dark cycle offered appropriate light exposure and darkness, promoting healthy larval growth.

For the second instar, temperature also had a significant impact on development (Figure 5A), with an overall average effect size of −0.8725 (CI: −0.9938; −0.7512; Figure S6, Table 2). The optimal growth temperature was 31 °C (Figure 5C). Compared to the first instar, the second instar larvae were more adapted to slightly lower temperatures, and a humidity level of 76% supported normal development. The ideal light cycle was 12:12 (12 h light:12 h dark), which provided the appropriate diurnal variation necessary for optimal growth and development during this stage.

Temperature had a similarly significant effect on the development of the third instar (Figure 5A), with an overall average effect size of −0.8763 (CI: −1.0132; −0.7393; Figure S7, Table 2). The optimal temperature for the third instar was 30 °C (Figure 5D). Compared to the previous instars, the temperature was slightly lower, and the humidity increased to 83%, which contributed to faster growth and higher survival rates. The light cycle remained at 12:12, continuing to play a crucial role in promoting development. At this stage, larvae exhibited a significantly higher demand for humidity, and the temperature of 30 °C proved to be the most suitable for their growth, providing the optimal developmental conditions.

In the fourth instar, temperature remained a critical factor influencing development (Figure 5E), with an overall average effect size of −0.72 (CI: −0.9332; −0.5068; Figure S8, Table 2). Within the temperature range of 17–34 °C, the optimal growth temperature was 31 °C (Figure 5F). At this stage, the temperature slightly increased to 31 °C, and the humidity decreased slightly to 82%, yet remained at a relatively high level. Compared to the third instar, the fourth instar larvae demonstrated greater adaptability to the 12:12 light: dark cycle, which continued to be the optimal photoperiod for growth and development. The combination of temperature and humidity provided ideal developmental conditions, ensuring efficient growth.

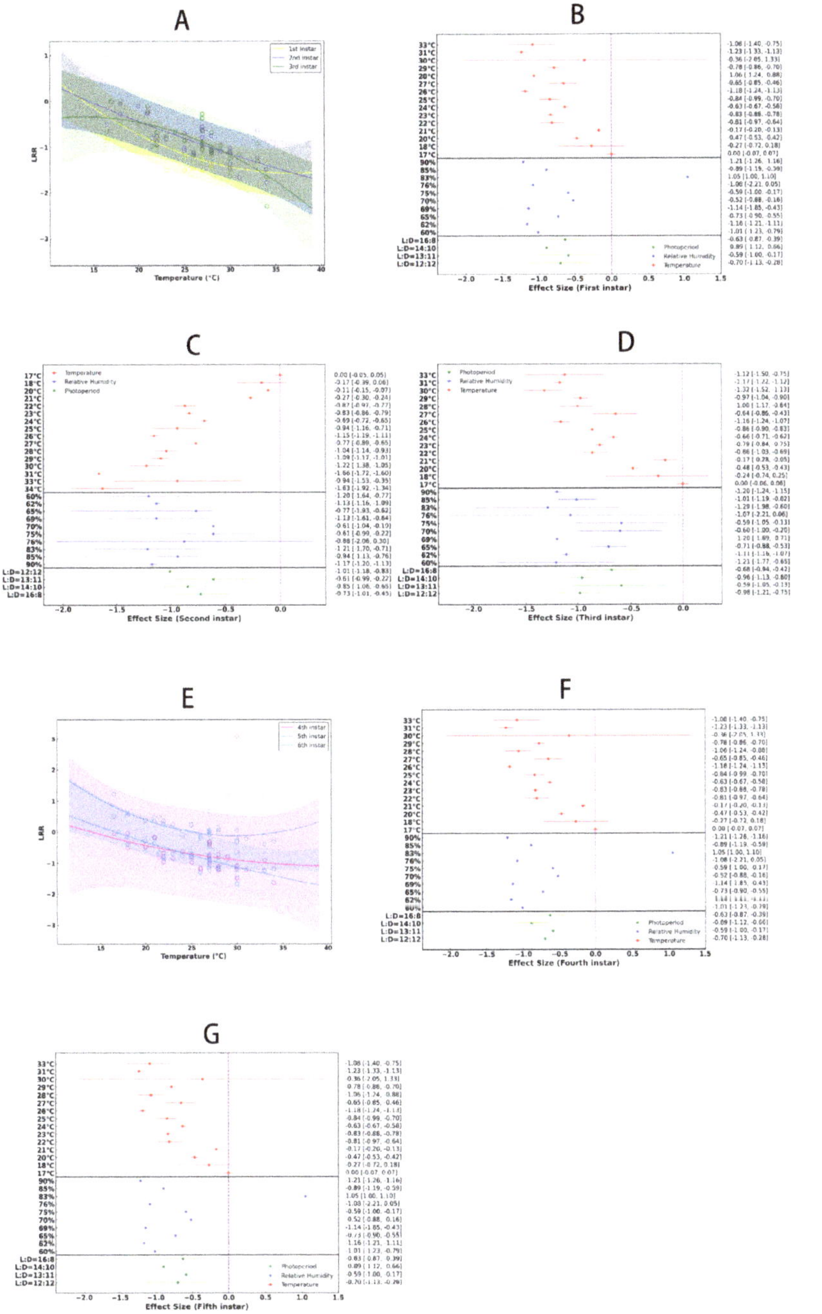

Figure 5. The impact of temperature on various developmental stages of *S. litura* is demonstrated. Panel (**A**) shows how the developmental time of first-to-third instar larvae changes with temperature variations, while Panels (**B∓D**) depict their response to external environmental conditions. Similarly, Panel (**E**) illustrates the changes in developmental time of fourth-to-sixth instar larvae with temperature variations, and Panels (**F**,**G**) present the response of fourth-to-fifth instar larvae to external environmental factors.

In the case of the fifth instar, temperature increase resulted in a shortened developmental period, with an overall average effect size of −0.8143 (CI: −0.9701; −0.6586; Figure S9, Table 2). Within the 17–34 °C temperature range, the developmental time at the fifth instar stage significantly decreased with rising temperature (Figure 5E). Different humidity and photoperiod conditions affected the developmental rate of the fifth instar. The optimal conditions for the fifth instar were found at 30 °C, 76% relative humidity, and a photoperiod of 12:12 (L:D) (Figure 5G).

However, no significant effect of temperature increase was observed for the sixth instar stage, with an overall average effect size of 0.0597 (CI: −0.1070; 0.2263; Figure S10, Table 2). Within the temperature range of 17–34 °C, the developmental time for the sixth instar was not significantly affected by temperature. The cumulative effect sizes suggest that within this temperature range, temperature changes did not significantly influence the developmental time at the sixth instar stage.

3.5. The Effect of Temperature Changes on the Lifespan of Adult *S. litura*

Temperature significantly affects the lifespan of female *S. litura* (Figure 6A). The shortest lifespan for female adults occurs under conditions of 33 °C temperature, 65% humidity, and a 12:12 light cycle (Figure 6B). The temperature response curve indicates that at this temperature, the lifespan of female adults is significantly minimized. The overall average effect size is −0.9734 (CI: −1.0667; −0.8801; Figure S11, Table 2), suggesting that an increase in temperature significantly reduces the lifespan of female adults.

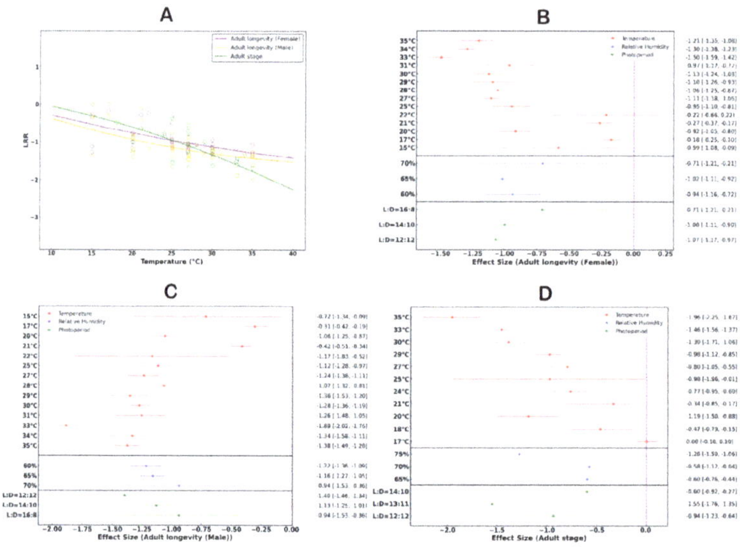

Figure 6. The effects of temperature on the lifespan of adult *S. litura*. (**A**) shows that the changes in the lifespan of female and male adults with increasing temperature. (**B**) depicts the response of female adult lifespan to variations in external environmental conditions. (**C**) depicts the response of male adult lifespan to variations in external environmental conditions. (**D**) depicts the response of the adult stage duration to variations in external environmental conditions.

Similarly, temperature also has a significant impact on the lifespan of male *S. litura* adults (Figure 7A). The shortest lifespan for male adults occurs under conditions of 33 °C temperature, 60% humidity, and a 12:12 light cycle (Figure 6C). According to this temperature response curve, the lifespan of male adults reaches its minimum under these conditions. The overall average effect size is −1.1394 (CI: −1.2488; −1.03; Figure S12,

Table 2), indicating that elevated temperatures have a significant negative effect on the lifespan of male adults.

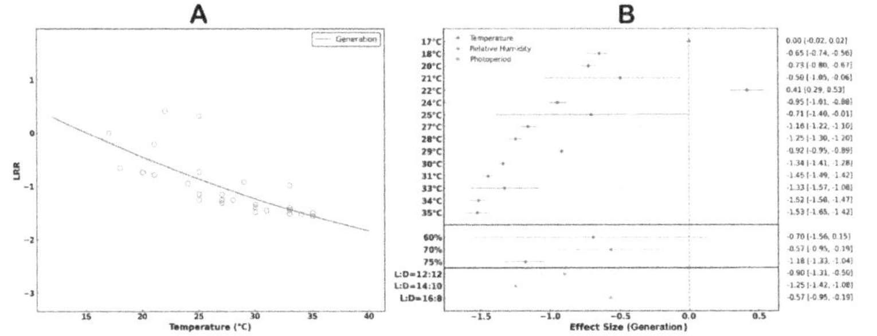

Figure 7. The influence of temperature variation on the life cycle of *S. litura*. (**A**) illustrates how the life cycle of *S. litura* changes with increasing temperature. (**B**) shows the response of the life cycle of *S. litura* to variations in external environmental conditions.

Additionally, the studies indicate that an increase in temperature shortens the adult stage duration of *S. litura*, with an overall average effect size of −1.0392 (CI: −1.2611; −0.8173; Figure S13, Table 2). Within the temperature range of 17–35 °C, the adult stage duration significantly decreases as temperature rises (Figure 6A). The results show that the shortest adult stage duration occurs when the temperature reaches 35 °C, with relative humidity at 75%, and a light cycle of 13:11 (Figure 6D).

3.6. The Effect of Temperature on the Generation Cycle of *S. litura*

The studies indicate that an increase in temperature shortens the generation cycle of *S. litura*, with an overall average effect size of −0.9917 (CI: −1.1961; −0.7873; Figure S14). Within the temperature range of 17–35 °C, the generation cycle time significantly decreases as temperature rises (Figure 7A). The findings suggest that variations in the light cycle have a significant effect on the generation cycle time. The fastest generation cycle is observed when the temperature reaches 35 °C, with relative humidity at 75%, and a light cycle of 14:10 (Figure 7B).

3.7. Model Validation

Funnel plots and radar charts were utilized to assess whether publication bias influenced the results, and the failsafe number was calculated to verify the reliability of the findings. The results demonstrate that the funnel plot (Figure 8A, $z = 11.7603$, $p = 0.0702$), radar chart (Figure 8B), and failsafe number ($n = 276385$) all confirm the robustness of our results.

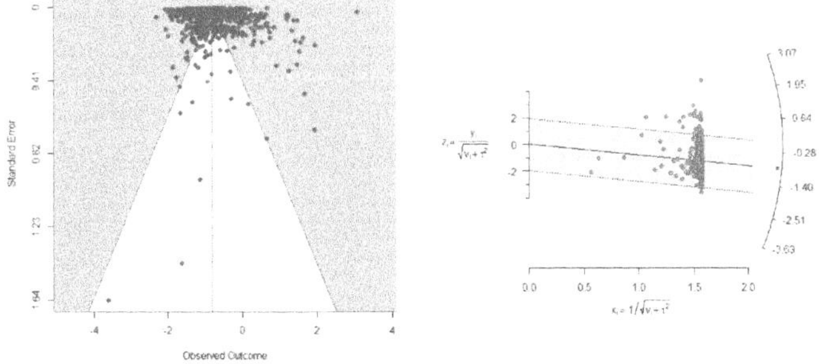

Figure 8. Funnel chart and radar chart. (**A**) shows a funnel chart, and (**B**) shows a radar chart.

4. Discussion

As ectothermic organisms, insect populations are heavily influenced by temperature [33]. Global warming has been shown to significantly alter the spatial distribution of numerous insect species, leading to range expansion and regional shifts [34]. Temperature is thus a critical factor in the survival, development, and reproduction of *S. litura*. This study confirms that temperature profoundly impacts the growth and developmental cycle of this pest. Within an optimal temperature range, increasing temperatures significantly accelerate its growth and development. An analysis was conducted on the relationship between the developmental stages of *S. litura* and temperature, resulting in feedback curves that highlight temperature-dependent variations across all stages (Table 3). The findings indicate that *S. litura* thrives best at temperatures between 30–35 °C and relative humidity levels of 60–83%. Global warming is projected to drastically alter ecosystems, reshaping habitat structures and increasing insect prevalence and distribution, which may exacerbate agricultural damage. This investigation points out that under anticipated warming conditions, *S. litura* is likely to develop stronger adaptive capabilities, posing greater challenges to agricultural production.

Table 3. The optimal environmental conditions for each developmental stage of *S. litura*.

Developmental History	Ideal Survival Temperature	Ideal Relative Humidity for Survival	Optimal Light/Dark Duration for Survival
1st instar	33 °C	76%	L:D = 14:10
2nd instar	31 °C	76%	L:D = 12:12
3ird instar	30 °C	83%	L:D = 12:12
4th instar	31 °C	82%	L:D = 12:12
5th instar	30 °C	76%	L:D = 12:12
Adult longevity (Female)	33 °C	65%	L:D = 12:12
Adult longevity (male)	33 °C	60%	L:D = 12:12
Adult stage	35 °C	75%	L:D = 13:11
Egg	34 °C	60%	L:D = 12:12
Egg to adult	33 °C	60%	L:D = 12:12
Generation	35 °C	75%	L:D = 14:10
Pre-pupa	30 °C	65%	L:D = 13:11
Pupa	30 °C	75%	L:D = 12:12
Fertility	34 °C	75%	L:D = 14:10
Larval stage	34 °C	75%	L:D = 12:12
Spodoptera litura	30−35 °C	60–83%	–

Temperature significantly influences the biological traits of *S. litura*. This meta-analysis indicates that increased temperatures accelerate both the developmental rate and reproductive potential of *S. litura*. However, the positive effects of temperature diminish or may even be reversed under extreme heat conditions. This finding aligns with previous studies [35], indicating that within an optimal temperature range, *S. litura* can shorten its life cycle and increase population size. Notably, when temperatures exceed their upper thermal tolerance (typically above 35 °C), both development rates and survival decrease significantly. This is likely due to the adverse physiological effects of temperature stress [36], suggesting that extreme heat may inhibit the expansion of *S. litura* populations. The responses to temperature vary across different developmental stages. According to the results of the meta-analysis, the development rates of the first to fifth larval instars increase as temperature rises, although sensitivity to temperature differs among instars. The early larval stages, particularly the 1st and 2nd instars, are highly sensitive to temperature fluctuations, showing a marked increase in development rates at higher temperatures. However, these stages also experience increased mortality at elevated temperatures, reflecting the high survival pressure during the larval phase. In contrast, the impact of temperature on the 6th instar is relatively moderate, suggesting that the later instar develops greater thermal tolerance. This result aligns with previous studies [35], demonstrating that later stages exhibit stronger adaptability to environmental changes. Temperature also significantly affects the oviposition behavior of female adults. According to the meta-analysis, oviposition rates rise with increasing temperatures, particularly within the ideal range of 30–35 °C, where female adults achieve peak reproductive output. However, when temperatures exceed 35 °C, oviposition rates decline, indicating that high temperatures suppress reproductive capacity. This may result from negative effects on ovarian development or egg quality, consistent with the hypothesis of impaired reproductive functions under thermal stress [37]. Thus, while moderate warming may promote population growth, extreme heat could suppress reproduction under global warming scenarios. The influence of temperature on the developmental duration of *S. litura* exhibits clear stage-specific effects. Developmental periods shorten with increasing temperature, highlighting the role of temperature in accelerating growth. However, this effect shows a lag during the adult stage. The meta-analysis reveals cumulative effects of temperature across developmental stages: while higher temperatures accelerate early-stage development, they also negatively impact adult survival and behavior. At elevated temperatures, the lifespan of adults is markedly reduced, and the generation cycle speeds up, likely because of the suppressive influence of heat on metabolic energy processes and behavioral capacity. Consequently, while temperature positively regulates developmental periods, high temperatures may limit adult survival, potentially affecting long-distance migration and dispersal.

In this study, we examined the impact of temperature variation on the growth and development of *S. litura*. However, in addition to temperature, environmental factors such as humidity and light also play crucial roles in the developmental processes of this species. Firstly, humidity is one of the key environmental factors influencing insect growth and development. Changes in humidity not only directly affect the physiological functions of *S. litura*, but can also indirectly impact its reproductive capacity and survival rate by altering the water balance within the insect. Studies have shown that higher humidity tends to increase egg-hatching rates and larval survival in *S. litura*, suggesting that an optimal humidity environment is beneficial for its development. Secondly, light is another important factor influencing the growth and development of *S. litura*. As a nocturnal insect, *S. litura* is primarily active at night and seeks shelter during the day. Variations in light intensity and the circadian rhythm can affect its activity patterns, foraging behavior, and physiological processes. With the ongoing global warming, it is anticipated that the distribution of *S. litura* will increase significantly [38,39]. In a warming world, will the adaptability of *S. litura* improve? To address this question, we reviewed and analyzed 17 relevant studies using meta-analysis to evaluate the responses of *S. litura* to temperature changes. The results indicated that within the temperature range of 15–38 °C, the adaptability of *S. litura* improved progressively with increasing temperatures. This finding aligns with previous studies, further validating the reliability of the model. Given the potential for *S. litura* to cause severe agricultural damage and economic losses, monitoring its adaptability is essential [40]. This study predicts, through meta-analysis, that *S. litura* exhibits peak adaptability at temperatures between 30–35 °C These findings provide critical insights for early warning systems regarding the direction of insect invasions and suitable habitats, offering valuable information for future monitoring and forecasting efforts.

This meta-analysis included a systematic review and selection of 17 studies investigating the effects of temperature on the biological traits of *S. litura*. The studies primarily focused on regions where *S. litura* is most prevalent, including East Asia, Southeast Asia, South Asia, and West Africa. Most of the research utilized controlled laboratory experiments, with a few employing field trials. While temperature is widely recognized as a critical environmental factor influencing *S. litura*, the studies exhibited heterogeneity in experimental conditions, species, and reporting methods. To address this variability, we employed a random-effects model and conducted subgroup analyses to explore the specific impacts of different temperatures on the biological traits of *S. litura*. Global warming is expected to profoundly alter ecosystem functions and structures, leading to shifts in the distribution of biological habitats [41,42]. The increasing prevalence and spread of insects, along with the resultant damages, will significantly impact agricultural production. This study highlights that under future warming scenarios, the adaptability of *S. litura* is likely to strengthen. The findings provide policymakers with critical insights to develop effective pest management strategies, thereby mitigating the potential widespread economic losses caused by pests under a warming climate.

5. Conclusions

This study conducted a systematic evaluation of the effects of temperature fluctuations on the biological traits of *S. litura* within the context of global climate change, using a meta-analysis. The results indicated that within the optimal temperature range of 30 °C to 35 °C, increasing temperatures significantly accelerated the developmental rate of *S. litura*, shortened its life cycle, and enhanced the oviposition rate of female adults. However, under extremely high-temperature conditions (above 35 °C), the developmental rate slowed, mortality increased, and the reproductive capacity of female adults was significantly suppressed, suggesting an upper-temperature threshold beyond which *S. litura*'s biological processes are negatively impacted. In addition to temperature, other environmental factors such as humidity and light also influence the growth and development of *S. litura*. The interaction between these factors and temperature, alongside the implications of climate change, presents a complex dynamic for *S. litura* populations. As global temperatures rise, shifts in humidity and light conditions may exacerbate or mitigate the impact of temperature extremes, affecting the overall population dynamics and adaptive strategies of this species. Importantly, the sensitivity of *S. litura* to temperature changes varies across developmental stages, with the larval stage being the most sensitive to high temperatures. This suggests that managing temperature, humidity, and light conditions will be crucial for predicting future pest outbreaks and for developing effective control strategies.

Supplementary Materials: The following supporting information can be downloaded at: https://www.mdpi.com/article/10.3390/insects16040355/s1. Figure S1: Comparison of effect sizes between random-effects and fixed-effects models; Figure S2: Comparison of effect sizes between random-effects and fixed-effects models; Figure S3: Comparison of effect sizes between random-effects and fixed-effects models; Figure S4: Comparison of effect sizes between random-effects and fixed-effects models; Figure S5: Comparison of effect sizes between random-effects and fixed-effects models; Figure S6: Comparison of effect sizes between random-effects and fixed-effects models; Figure S7: Comparison of effect sizes between random-effects and fixed-effects models; Figure S8: Comparison of effect sizes between random-effects and fixed-effects models; Figure S9: Comparison of effect sizes between random-effects and fixed-effects models; Figure S10: Comparison of effect sizes between random-effects and fixed-effects models; Figure S11: Comparison of effect sizes between random-effects and fixed-effects models; Figure S12: Comparison of effect sizes between random-effects and fixed-effects models; Figure S13: Comparison of effect sizes between random-effects and fixed-effects models; Figure S14: Comparison of effect sizes between random-effects and fixed-effects models;

Figure S15: Comparison of effect sizes between random-effects and fixed-effects models; Figure S16: Comparison of effect sizes between random-effects and fixed-effects models; Figure S17: Comparison of effect sizes between random-effects and fixed-effects models; Figure S18: Comparison of effect sizes between random-effects and fixed-effects models; Figure S19: Comparison of effect sizes between random-effects and fixed-effects models.

Author Contributions: Conceptualization, Z.Z.; methodology, Z.L.; software, Z.L. and H.Z.; formal analysis, C.F.; investigation, C.F. and T.G.; data curation, X.D.; writing—original draft preparation, Z.L. and H.Z.; writing—review and editing, D.X. and X.D.; supervision, Z.Z. and D.X. All authors have read and agreed to the published version of the manuscript.

Funding: This work was funded by the Fundamental Research Funds of China West Normal University (20A007, 20E051, 21E040 and 22kA011).

Data Availability Statement: The data supporting the results are available in a public repository at: https://doi.org/10.6084/m9.figshare.28587818, accessed on 11 March 2025.

Conflicts of Interest: The authors declare no conflicts of interest.

References

1. Haq, M.; Mia, M.T.; Rabbi, M.F.; Ali, M.A. Incidence and severity of rice diseases and insect pests in relation to climate change. In *Climate Change and Food Security in South Asia*; Springer: Berlin/Heidelberg, Germany, 2011; pp. 445–457.
2. Hakala, K.; Hannukkala, A.; Huusela-Veistola, E.; Jalli, M.; Peltonen-Sainio, P. Pests and diseases in a changing climate a major challenge for Finnish crop production. *Agric. Food Sci.* **2011**, *20*, 3–14. [CrossRef]
3. Skendžić, S.; Zovko, M.; Živković, I.P.; Lešić, V.; Lemić, D. The impact of climate change on agricultural insect pests. *Insects* **2021**, *12*, 440. [CrossRef] [PubMed]
4. Subedi, B.; Poudel, A.; Aryal, S. The impact of climate change on insect pest biology and ecology: Implications for pest management strategies, crop production, and food security. *J. Agric. Food Res.* **2023**, *14*, 100733.
5. Sharma, H.C. Climate change effects on insects: Implications for crop protection and food security. *J. Crop Improv.* **2014**, *28*, 229–259. [CrossRef]
6. Zakria, M.; Zaib, M.S.; Abbas, K.; Sarmad, M.; Zaka, S.M.; Noor-ul-Ane, M. Influence of temperature on the development and reproduction of *Spodoptera litura* (Fabricius) on castor bean: Implications for its use as a trap crop. *Arthropod-Plant Interact.* **2022**, *16*, 505–515.
7. Ahmad, M.; Arif, M.I.; Ahmad, M. Occurrence of insecticide resistance in field populations of *Spodoptera litura* (Lepidoptera: Noctuidae) in Pakistan. *Crop Prot.* **2007**, *26*, 809–817.
8. Ahmad, M.; Ghaffar, A., Rafiq, M. Host plants of leaf worm, *Spodoptera litura* (Fabricius) (Lepidoptera: Noctuidae) in Pakistan. *Asian J. Agric. Biol.* **2013**, *1*, 23–28.
9. Lu, Y.; Yuan, M.; Gao, X.; Kang, T.; Zhan, S.; Wan, H.; Li, J. Identification and validation of reference genes for gene expression analysis using quantitative PCR in *Spodoptera litura* (Lepidoptera: Noctuidae). *PLoS ONE* **2013**, *8*, e68059.
10. Yoon, S.; Lee, W. Methodological analysis of bioclimatic variable selection in species distribution modeling with application to agricultural pests (*Metcalfa pruinosa* and *Spodoptera litura*). *Comput. Electron. Agric.* **2021**, *190*, 106430.
11. Maharjan, R.; Hong, S.; Ahn, J.; Yoon, Y.; Jang, Y.; Kim, J.; Lee, M.; Park, K.; Yi, H. Temperature and host plant impacts on the development of *Spodoptera litura* (Fabricius)(Lepidoptera: Noctuidae): Linear and nonlinear modeling. *Insects* **2023**, *14*, 412. [CrossRef]
12. Miyashita, K. Effects of constant and alternating temperatures on the development of *Spodoptera litura* F.: Lepidoptera: Noctuidae. *Appl. Entomol. Zool.* **1971**, *6*, 105–111. [CrossRef]
13. Rao, M.S.; Prasad, T.V. Temperature based phenology model for predicting establishment and survival of *Spodoptera litura* (Fab.) on groundnut during climate change scenario in India. *J. Agrometeorol.* **2020**, *22*, 24–32.
14. Zhong, T.; Gong, L.; Pan, Y.; Li, J.; Lu, A.; Liu, L.; Wu, H.; Zhao, Z.; Wang, L. Performance of *Spodoptera litura* (Lepidoptera: Noctuidae) in responses to different amplitudes of alternating temperatures across permissive warm temperature regimes. *J. Econ. Entomol.* **2024**, *117*, 1041–1046. [CrossRef] [PubMed]
15. Zhu, S.; Lu, Z.; Chen, L.; Yu, W.; Zhang, S. Effect of temperature and food on *Spodoptera litura* population. *Ying Yong Sheng Tai Xue Bao J. Appl. Ecol.* **2000**, *11*, 111–114.
16. Pham, T.A.; Hwang, S. High temperatures reduce nutrients and defense compounds against generalist *Spodoptera litura* F. in Rorippa dubia. *Arthropod-Plant Interact.* **2020**, *14*, 333–344. [CrossRef]

17. Karmakar, P.; Pal, S. Influence of temperature on food consumption and utilization parameters of the common cutworm, *Spodoptera litura* (Fab)(Lepidoptera: Noctuidae). *J. Entomol. Zool. Stud.* **2017**, *5*, 92–95.
18. Harvey, J.A.; Tougeron, K.; Gols, R.; Heinen, R.; Abarca, M.; Abram, P.K.; Basset, Y.; Berg, M.; Boggs, C.; Brodeur, J. Scientists' warning on climate change and insects. *Ecol. Monogr.* **2023**, *93*, e1553. [CrossRef]
19. Hedges, L.V.; Gurevitch, J.; Curtis, P.S. The meta-analysis of response ratios in experimental ecology. *Ecology* **1999**, *80*, 1150–1156. [CrossRef]
20. Cooper, C.; Booth, A.; Varley-Campbell, J.; Britten, N.; Garside, R. Defining the process to literature searching in systematic reviews: A literature review of guidance and supporting studies. *Bmc Med. Res. Methodol.* **2018**, *18*, 85. [CrossRef]
21. Dinet, J.; Favart, M.; Passerault, J.M. Searching for information in an online public access catalogue (OPAC): The impacts of information search expertise on the use of Boolean operators. *J. Comput. Assist. Learn.* **2004**, *20*, 338–346. [CrossRef]
22. Viechtbauer, W. Conducting meta-analyses in R with the metafor package. *J. Stat. Softw.* **2010**, *36*, 1–48.
23. Borenstein, M.; Hedges, L.V.; Higgins, J.P.; Rothstein, H.R. A basic introduction to fixed-effect and random-effects models for meta-analysis. *Res. Synth. Methods* **2010**, *1*, 97–111. [PubMed]
24. Corbeil, R.R.; Searle, S.R. Restricted maximum likelihood (REML) estimation of variance components in the mixed model. *Technometrics* **1976**, *18*, 31–38. [CrossRef]
25. Kosimov, Z.O. Methods for estimating integrated variance: Jump robustness issues in high frequency time series. *Econ. Math. Methods* **2024**, *60*, 107–117. [CrossRef]
26. Garamszegi, L.Z.; Markó, G.; Herczeg, G. A meta-analysis of correlated behaviours with implications for behavioural syndromes: Mean effect size, publication bias, phylogenetic effects and the role of mediator variables. *Evol. Ecol.* **2012**, *26*, 1213–1235.
27. Clark, T.S.; Linzer, D.A. Should I use fixed or random effects? *Political Sci. Res. Methods* **2015**, *3*, 399–408.
28. Lajeunesse, M.J. Bias and correction for the log response ratio in ecological meta-analysis. *Ecology* **2015**, *96*, 2056–2063. [PubMed]
29. Higgins, J.P.; Thompson, S.G. Quantifying heterogeneity in a meta-analysis. *Stat. Med.* **2002**, *21*, 1539–1558.
30. Thornton, A.; Lee, P. Publication bias in meta-analysis: Its causes and consequences. *J. Clin. Epidemiol.* **2000**, *53*, 207–216.
31. Peters, J.L.; Sutton, A.J.; Jones, D.R.; Abrams, K.R.; Rushton, L. Comparison of two methods to detect publication bias in meta-analysis. *JAMA* **2006**, *295*, 676–680. [CrossRef]
32. Duval, S.; Tweedie, R. A nonparametric "trim and fill" method of accounting for publication bias in meta-analysis. *J. Am. Stat. Assoc.* **2000**, *95*, 89–98.
33. Robinet, C.; Roques, A. Direct impacts of recent climate warming on insect populations. *Integr. Zool.* **2010**, *5*, 132–142.
34. Lehmann, P.; Ammunét, T.; Barton, M.; Battisti, A.; Eigenbrode, S.D.; Jepsen, J.U.; Kalinkat, G.; Neuvonen, S.; Niemelä, P.; Terblanche, J.S. Complex responses of global insect pests to climate warming. *Front. Ecol. Environ.* **2020**, *18*, 141–150.
35. Rao, M.S.; Manimanjari, D.; Rao, A.C.R.; Swathi, P.; Maheswari, M. Effect of climate change on *Spodoptera litura* Fab. on peanut: A life table approach. *Crop Prot.* **2014**, *66*, 98–106.
36. Neven, L.G. Physiological responses of insects to heat. *Postharvest Biol. Technol.* **2000**, *21*, 103–111.
37. Régnière, J.; Powell, J.; Bentz, B.; Nealis, V. Effects of temperature on development, survival and reproduction of insects: Experimental design, data analysis and modeling. *J. Insect Physiol.* **2012**, *58*, 634–647.
38. Jung, J.M.; Byeon, D.H.; Jung, S.; Lee, W.H. Effect of climate change on the potential distribution of the common cutworm (*Spodoptera litura*) in South Korea. *Entomol. Res.* **2019**, *49*, 519–528. [CrossRef]
39. Fu, X.; Zhao, X.; Xie, B.; Ali, A.; Wu, K. Seasonal pattern of *Spodoptera litura* (Lepidoptera: Noctuidae) migration across the Bohai Strait in northern China. *J. Econ. Entomol.* **2015**, *108*, 525–538. [CrossRef]
40. Sahu, B.; Pachori, R.; Navya, R.N.; Patidar, S. Extent of damage by *Spodoptera litura* on cabbage. *J. Entomol. Zool. Stud.* **2020**, *8*, 1153–1156.
41. Sangle, P.M.; Satpute, S.B.; Khan, F.S.; Rode, N.S. Impact of climate change on insects. *Trends Biosci.* **2015**, *8*, 3579–3582.
42. Jaworski, T.; Hilszczanski, J. The effect of temperature and humidity changes on insects development their impact on forest ecosystems in the expected climate change. *Lesn. Pr. Badaw.* **2013**, *74*, 345. [CrossRef]

insects

MDPI

Communication

Hydrogen Stable Isotopes Indicate Reverse Migration of Fall Armyworm in North America

Eduardo S. Calixto and Silvana V. Paula-Moraes *

Entomology and Nematology Department, West Florida Research and Education Center, University of Florida, Jay, FL 32565, USA; calixtos.edu@gmail.com
* Correspondence: paula.moraes@ufl.edu

Simple Summary: Fall armyworm (*Spodoptera frugiperda*) is a destructive pest that causes significant crop damage, especially in the U.S., and has spread to various countries around the world. Understanding how these pests migrate is crucial for predicting outbreaks and developing effective management programs. In this study, we estimated the movement of fall armyworm moths by analyzing hydrogen isotopes in 324 samples collected at the edge of continental U.S., which is considered an interbreeding zone for this species. Our results indicate that fall armyworm moths migrate southward from northern U.S. regions, like the Corn Belt, including states such as Nebraska, South Dakota, Minnesota, Kansas and Wisconsin. This discovery provides important insights into the movement of this pest and potential spread of resistance alleles, which can help improve integrated pest management and insect resistance management. The findings are valuable for developing more targeted and timely pest management strategies to protect agriculture and ensure food security.

Abstract: Fall armyworm (FAW), *Spodoptera frugiperda* (J. E. Smith, 1797) (Lepidoptera: Noctuidae), is a major pest in the U.S. and has spread globally, causing severe agricultural losses in different countries. Due to its high mobility and potential for long-distance dispersal, understanding FAW migration is a key tool for forecasting outbreaks and implementing timely management measures. Recent studies using stable hydrogen isotopes indicated reverse (southward) migration of *Helicoverpa zea* Boddie (Lepidoptera: Noctuidae). Here, we tested the reverse migration hypothesis for FAW in North America. Estimation of the hydrogen isotopic ratio on 324 samples collected in Florida, an intermixing zone at the edge of the continental U.S., indicated evidence of reverse migration in samples of FAW moths. They showed a high probability of origin from the U.S. Corn Belt, with a greater probability of origin in Nebraska, South Dakota, Minnesota, Kansas and Wisconsin. This southward movement provides new insights into the risk of spreading pesticide resistance alleles in this species to southern regions and contributes to the improvement of integrated pest management and insect resistance management programs.

Keywords: *Spodoptera frugiperda*; migration; insect resistance risk; biogeochemical marker; dispersal; integrated pest management

1. Introduction

Fall armyworm (FAW), *Spodoptera frugiperda* (J.E. Smith, 1797) (Lepidoptera: Noctuidae), is one of the most economically significant pests in the United States. This highly polyphagous insect can feed on more than 350 plant species, showing a particular preference for crops in the family Poaceae, including corn, sorghum, rice, and turfgrass [1]. In 2016, FAW emerged as an invasive pest in regions outside its native range, spreading across multiple continents and causing substantial agricultural losses in the eastern hemisphere [2]. Its sporadic and highly unpredictable outbreaks [3], along with its aggressive feeding behavior, make FAW a key pest of major crops worldwide. Understanding the ecology and migratory patterns of FAW is paramount for forecasting seasonal movements and potential outbreaks, which can aid in developing timely and effective management strategies.

FAW is a highly mobile pest capable of dispersing long distances, with reports of 1600 km in 30 h [4]. Studies have shown that FAW moths migrate northward during the crop season, extending into the northern United States and southern Canada [5]. Since FAW does not enter diapause, two main hypotheses have been proposed to explain its population dynamics. The first, known as the "pied piper" migration hypothesis, suggests a northward movement during spring without a return to southern regions [6]. The second hypothesis, known as reverse (southward) migration, posits a southward movement where moths return to overwintering sites in southern Florida and Texas at the end of the crop season [3,5,7,8]. Although studies have shown support for a northward movement from overwintering regions to northern areas [3,5,7–9], there is a lack of supporting data for FAW movement from the northern regions of the U.S. and southern Canada to lower latitudes in North America [10,11].

Evidence for reverse migration has been hypothesized for corn earworm, *Helicoverpa zea* (Boddie), another economic noctuid pest in the U.S. [12,13]. Recent work using stable hydrogen isotopic ratios estimated moths migrating southward as far as the Caribbean basin [14]. This reverse migration phenomenon in *H. zea* together with previous studies on FAW migration suggests that a similar pattern might exist in FAW. For instance, the "pied piper" hypothesis proposes that FAW spreads northward each summer but cannot overwinter [7]; migration could not persist without a return movement, as there would be a potential loss of migratory genes in successive generations [15]; and selection could maintain migration if some individuals return southward [16]. These studies support the possibility of FAW reverse migration, which could have significant implications for integrated pest management (IPM) and insect resistance management (IRM) programs due to the extended persistence of resistance alleles to management tools, such as insecticides and transgenic Bt plants, at a continental scale. Here, we tested the hypothesis of potential reverse migration of FAW in North America by using stable hydrogen isotopes.

2. Materials and Methods

2.1. Data Collection

A long-term year-round moth trapping program using delta traps (Trécé, Inc., Adair, OK, USA) at the West Florida Research and Education Center, Jay, FL, USA (longitude: −87.143891, latitude: 30.773188) has been established since 2017. This trapping location is considered an intermixing zone at the edge of the continental U.S., where FAW moths from populations in South Florida and South Texas converge during migration [8,9]. The collected moths were stored at under −20 °C. A total of 324 samples of FAW moths were selected for this study, corresponding to about 56 moths per year (Table S2). Out of these, 19 moths were collected in April, 20 in May, 13 in June, 42 in July, 58 in August, 84 in September, 73 in October, and 15 in November.

2.2. Stable Hydrogen Isotopes Analysis

Each moth had the right forewing removed, prepared, and submitted to hydrogen isotopic ratio analyses, as described in Paula-Moraes et al. [14]. The method involves removing scales from the wings using a painting brush and cleaning the wing first with Goo and Adhesive Remover Spray Gel (Goo Gone, CC Holdings, Inc., New York, NY, USA) to eliminate any glue from sticky strips. Then, the wing was submerged in 70% ethanol for 24 h to remove any residual oils on its surface (Method S1). In the lab, this cleaning method produced the same results as the traditional 2:1 chloroform/methanol treatment (Method S1, [17]). When dried, wings were cut into small pieces (range: 0.084–0.191 mg, Table S2) and submitted to hydrogen isotope analyzes in the Stable Isotope Mass Spec Lab, University of Florida, Gainesville, FL, USA. Samples and standards were analyzed in a Thermo Electron DeltaV Plus isotope ratio mass spectrometer coupled with a ConFlo IV interface linked to a TCEA (high-temperature conversion elemental analyzer). After weighing and loading the samples and standards into 4 mm × 6 mm silver capsules, samples were left in 96-well plates for 48 h. This process ensures that isotopic composition of the samples and standards remained comparable and consistent [18]. To determine nonexchangeable hydrogen, two keratin standards (Caribou Hoof Standard—CBS, and Kudu Horn Standard—KHS) were used.

After placing the capsules into a zero Blank autosampler at 1400 °C, hydrogen isotopic values (δ^2H) were measured in a Picarro L2120-I isotopic liquid water and water vapor analyzer (Santa Clara, CA, USA) coupled with a Picarro A0211 high precision vaporizer and a CTC HTS PAL autosampler (Santa Clara, CA, USA). Precision was based on USGS42 = 2.99 ‰ (N = 11). To standardize the results, two internal University of Florida water standards (UW Antarctic water and Lake Tulane water) were used, which were calibrated using international standards (USGS49 and USGS50). Isotope results are reported in standard delta notation relative to Vienna Standard Mean Ocean Water. Detailed information of all the process can be found at Paula-Moraes et al. [14].

2.3. Inferences on the Probability of Origin

Probability of origin of each sample was inferred based on the methodology described in Ma et al. [19] using the package assignR [19] in the R software version 4.3.1 [20]. First, we built an isoscape based on the hydrogen isotope values of amount-weighted, growing-season precipitation at 5 arc-minute resolution. The isoscape was then calibrated using known values of hydrogen isotopes retrieved from published literature that were included in the assignR package [19]. The package contains a database with hydrogen isotope data from known-origin samples (sample values of *Danaus Plexippus*, monarch butterfly, already available in the R package). Then, calibration was performed to transform all hydrogen isotope data onto a common reference scale, ensuring comparability across datasets from different laboratories. We used the isotope values from the monarch butterfly database, a Lepidopteran species, to provide a realistic comparison of hydrogen isotope assimilation from the environment into wing tissues, ensuring consistency in taxon, geographic region, and analytical approach [21]. Then, a linear model between the environmental (precipitation) isoscape values and the known values (monarch butterfly values) was fitted to produce a calibrated isoscape. With the calibrated isoscape, we generated posterior probability maps for each unknown sample (i.e., field-collected samples) using the Bayesian inversion method. This approach calculates the scaled probability of origin for each grid cell and produces a raster map. Each cell's value represents the probability that it is the actual origin of the sample among all cells in the map. Finally, we calculated the average distance and direction based on the potential probabilities of origin. Based on the probability of origin maps, low hydrogen isotope ratios, the month of collection, and moth behavior and landscape, we identified moths likely migrating from northern locations.

3. Results

Most of the moths collected showed a high probability of originating in Florida, Texas, and the Caribbean region (see example of a sample in Figure 1). Four samples collected between August and October, the crop season in the northern U.S., had a high probability of a northern U.S. origin (Figure 2). The average distance flown by these moths based on the probability-weighted distance ranged from 1312 to 1897 km (Figure 3). The average bearing ranged from 104 to 140 degrees, potentially indicating a movement toward the southeast from areas with the highest probability of origin (Figure 4). This suggests a potential movement from north Texas and Oklahoma to upper Midwest and Corn Belt region, such as North Dakota, South Dakota, Minnesota, Nebraska, Iowa, and Kansas. Other regions, like the northeast and parts of the southern U.S., also showed some probability of origin (Figure 2).

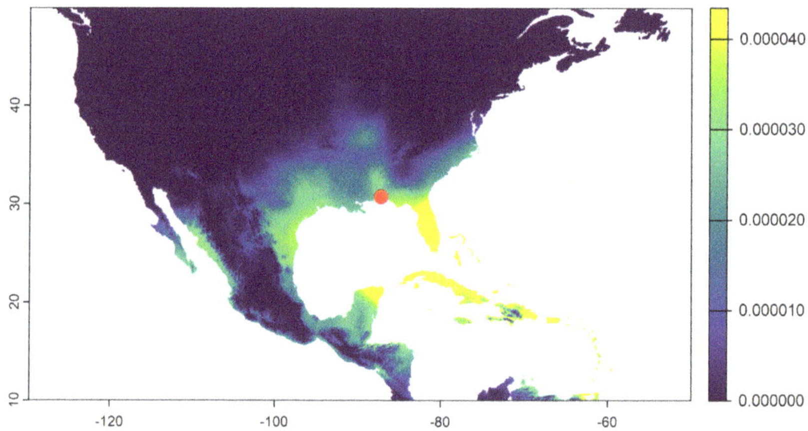

Figure 1. Example probability of origin map of a moth based on hydrogen isotope ratios with high probability of origin in Florida, Texas, and the Caribbean. The yellow color represents a higher probability of origin based on the posterior probability surface of a FAW moth collected at the edge of the continental U.S. See Table S2. Red dot shows where FAW moths were collected (WFREC, Jay, FL, USA).

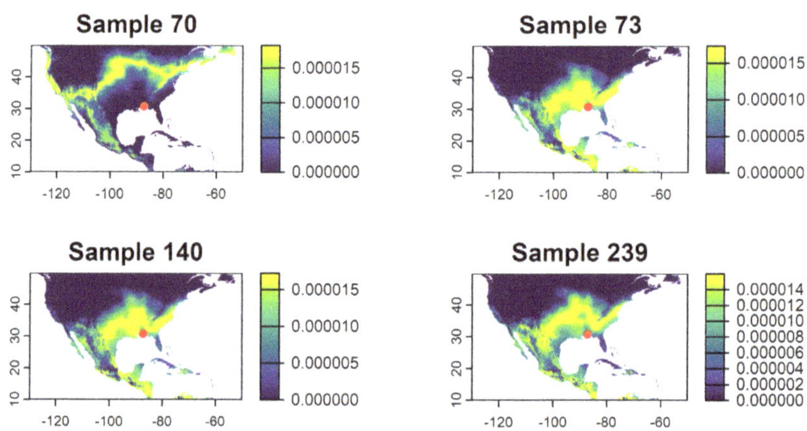

Figure 2. Evidence of FAW reverse migration based on hydrogen isotope ratios (δ^2H). The yellow color represents higher probability of origin based on the posterior probability surface of a FAW moth collected at the edge of the continental U.S. See Table S2. Red dots show where FAW moths were collected (WFREC, Jay, FL, USA).

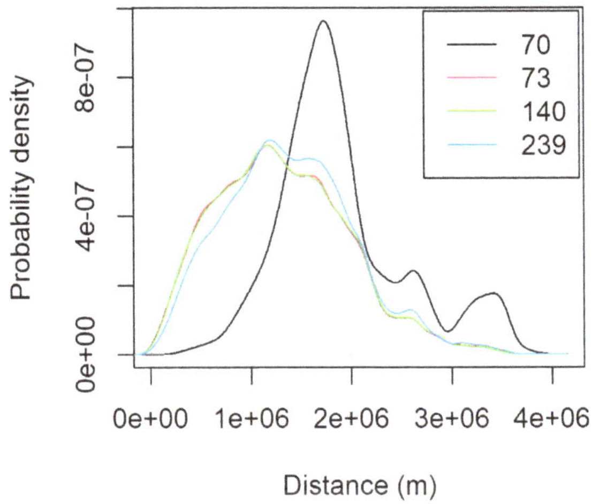

Figure 3. Probability density of the dispersal distance of each moth with a high probability of origin in the northern region of the U.S. based on the posterior probabilities calculated from Figure 2.

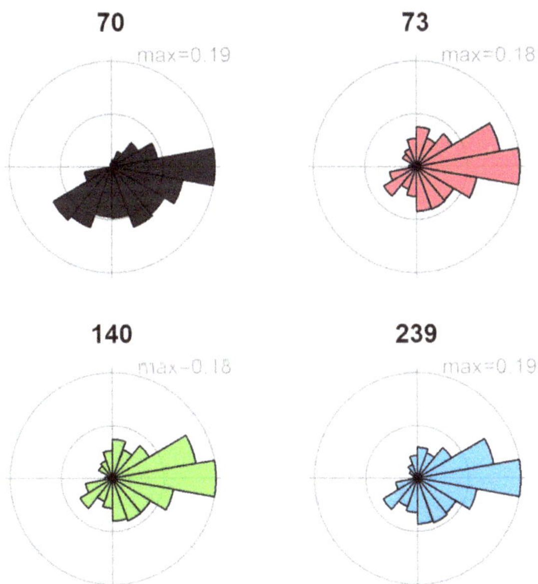

Figure 4. Likely dispersal direction of each moth with a high probability of origin in the northern region of U.S. based on the posterior probabilities calculated from Figure 2.

4. Discussion

Our results add new evidence for the reverse migration of FAW in North America [5,10,11], and this is the first study to support a large-scale southward movement of FAW using a biogeochemical marker. Based on the hydrogen isotopes ratio analysis, we identified a high probability of origin of moths in the northern region of the U.S., with a potential origin in the Corn Belt, the upper regions of southern states, and the northeast. Molecular studies have supported the northward movement of FAW due to the genetic similarity between populations in the northern and southern U.S. associated with prevailing wind patterns [5,8]. However, limited research has provided compelling evidence for the reverse migration at the end of the crop season [10,11]. Stable hydrogen isotope analysis of FAW wings offers valuable information on the geographical origins of larval feeding, shedding light on the migratory dynamics of this pest. This southward movement was already expected due to the lack of studies supporting the "pied piper" hypothesis, the potential loss of migratory genes present in successive populations migrating northward, and the evidence for reverse migration in other noctuid species in the U.S., particularly *H. zea* [12–14].

The probability maps suggest potential moth origins in the Corn Belt, upper regions of the southern states, the northeast, and some parts of the west. The rose diagrams and kernel density functions for distance provide valuable insights into migration direction and distance. However, these results must be interpreted alongside the species' ecology and landscape characteristics. Few studies have examined FAW migration across the Rocky Mountains (see [8]), where our δ^2H-based analysis also indicates a high probability of origin. Given the known ecological constraints and landscape barriers, we consider western origins unlikely despite some rose diagrams and probability maps suggesting movement from that direction. Instead, the strong overlap between the high probability of origin and the Corn Belt, combined with the timing of moth collection aligning with the crop season in that region, provides robust support for a reverse (southward) migration pattern.

Out of the 324 moths used in our analyses, 88 were collected during October and November from 2017 to 2023 (Table S2), with all other moths collected from April to September in the same period (Table S2). Southward movement potentially occurs at the end of the crop season in the northern states, usually around October–November. The low number of moths selected from the end of the crop season used in our study limits the probability of identifying more moths migrating from the northern region. We expected an increase in the number of moths toward the end of the year as the host crops decline and temperatures drop in the northern region. Since we used a broad range of collection months to test our hypothesis, we could evaluate when moths begin migrating southward. The moth with the highest probability of reverse migration based on both probability maps and collection period was sample 239, which was collected in October. Future studies should continue testing the reverse migration of FAW using samples collected, as far as possible, during the end of the crop season in the southern U.S., while also including some samples from earlier months.

One primary limitation of our study is the uncertainty about the very precise origin of migrating FAW moths. While our isotope analysis provides strong evidence for reverse migration, multidisciplinary approaches integrating molecular analyses and additional stable isotopes could enhance our understanding of migration patterns by providing finer resolution on natal origins and dispersal routes [5,8,9,22]. Additionally, although FAW is a polyphagous pest, it has a strong preference for Poaceae, particularly crops such as corn, sweet corn, sorghum, rice, and turfgrass [1]. Since these host plants belong to the same family, and the average fractionations between host plant and lepidopteran moths are around 3‰ [17], the variability in δ^2H due to plant fractionation is expected to be relatively low among host species. Finally, agricultural irrigation could introduce additional variation in δ^2H values compared to precipitation-based isoscapes. Although only about 17% of U.S. corn, the primary host plant of FAW, is irrigated [23], but water sources such as groundwater or reservoirs with high evaporation rates could influence δ^2H signatures. Nonetheless, these effects are likely to be minimal as groundwater and reservoir water typically mix with precipitation over time, reducing localized deviations [24]. These factors can influence δ^2H signatures, but the broad latitudinal δ^2H gradient used in our study mitigates the impact of small-scale variations as any minor isotopic shifts are likely overshadowed by the larger geographic patterns [14].

Understanding the movement of FAW moths is a key tool for IPM programs due to their high mobility, migratory potential, and polyphagous feeding habits. Historically, challenges related to the management of FAW are related to the occurrence of outbreaks, which are usually sporadic and unpredictable [3]. Although monitoring systems have advanced in the detection of moths and provided some insights into potential population dynamics of moth pests, e.g., [25–27], it is still important to refine forecasting methods and identify migratory routes to enable the timely adoption of management strategies in IPM programs. In addition, this species has evolved resistance to pesticides. For instance, resistance to Bt toxins [28–30] and synthetic insecticides [31,32] has already been documented in FAW populations worldwide. Our results indicate that if resistance selection in northern FAW populations acts as a 'source' of resistance alleles, then mathematical models should account for evolution of resistance at a continental scale within IRM programs. Overall, our study provides important insights into the population dynamics of FAW in the U.S.

Supplementary Materials: The following supporting information can be downloaded at https://www.mdpi.com/article/10.3390/insects16050471/s1. Table S1: Description of treatments used to clean moth wings prior to stable isotope analysis, including presence of scales, solvent type, and cleaning method. All wings were air-dried after solvent application method. NS—No scales, GG—Goo Gone. Table S2: Hydrogen isotope ratio of 324 moths collected over 6 years during crop season at the West Florida Research and Education Center, Jay, FL, USA. Figure S1: Hydrogen isotopic ratios of the wings of *Helicoverpa zea* (a) and *Spodoptera frugiperda* (b) under different treatments used to clean wings prior to stable isotope analysis (see Table S1). (a) GLM: $\chi^2 = 98.8$, $p < 0.001$, (b) GLM: $\chi^2 = 45.9$, $p < 0.001$. Different letters represent significant difference based on Tukey's post hoc test ($p < 0.05$). NS—No scales, GG—Goo Gone. References [17,33] are cited in the supplementary materials.

Author Contributions: E.S.C.: Conceptualization; data curation; investigation; validation; formal analysis; visualization; writing—original draft. S.V.P.-M.: Conceptualization; data curation; investigation; validation; supervision; resources; project administration; funding acquisition; writing—original draft. All authors have read and agreed to the published version of the manuscript.

Funding: This research was partially funded by Florida Peanut checkoff money, Cotton Inc., and by the following funding from USDA NIFA—National Institute of Food and Agriculture: hatch project FLAENY-006203, Ecology and Management of Arthropods in Corn FLA-ENY-006003, 2642.

Data Availability Statement: The original contributions presented in this study are included in the article/Supplementary Materials. Further inquiries can be directed to the corresponding author.

Acknowledgments: We thank the members of the Entomology lab at the West Florida Research and Education Center—UF for their assistance with insect collection and preparation, and Jason Curtis for conducting the hydrogen isotope analysis.

Conflicts of Interest: The authors declare no conflicts of interest.

Abbreviations

The following abbreviations are used in this manuscript:

FAW	Fall armyworm
IPM	Integrated pest management
IRM	Insect resistance management

References

1. Montezano, D.G.; Sosa-Gómez, D.; Specht, A.; Roque-Specht, V.F.; Sousa-Silva, J.C.; de Paula-Moraes, S.; Peterson, J.A.; Hunt, T. Host plants of *Spodoptera frugiperda* (Lepidoptera: Noctuidae) in the Americas. *Afr. Entomol.* **2018**, *26*, 286–300. [CrossRef]

2. Goergen, G.; Kumar, P.L.; Sankung, S.B.; Togola, A.; Tamò, M. First Report of Outbreaks of the Fall Armyworm *Spodoptera frugiperda* (J E Smith) (Lepidoptera, Noctuidae), a New Alien Invasive Pest in West and Central Africa. *PLoS ONE* **2016**, *11*, e0165632. [CrossRef] [PubMed]
3. Johnson, S.J. Migration and the Life History Strategy of the Fall Armyworm, *Spodoptera frugiperda* in the Western Hemisphere. *Int. J. Trop. Insect Sci.* **1987**, *8*, 543–549. [CrossRef]
4. Rose, A.H.; Silversides, R.H.; Lindquist, O.H. Migration flight by an aphid, *Rhopalosiphum maidis* (Hemiptera: Aphididae), and a noctuid, *Spodoptera frugiperda* (Lepidoptera: Noctuidae). *Can. Entomol.* **1975**, *107*, 567–576. [CrossRef]
5. Nagoshi, R.N.; Meagher, R.L. Review of Fall Armyworm (Lepidoptera: Noctuidae) Genetic Complexity and Migration. *Fla. Entomol.* **2008**, *91*, 546–554. [CrossRef]
6. Dingle, H. Function of migration in the seasonal synchronization of insects. *Entomol. Exp. Appl.* **1982**, *31*, 36–48. [CrossRef]
7. Luginbill, P. *The Fall Army Worm*; US Department of Agriculture: Washington, DC, USA, 1928.
8. Tessnow, A.E.; Nagoshi, R.N.; Meagher, R.L.; Fleischer, S.J. Revisiting fall armyworm population movement in the United States and Canada. *Front. Insect Sci.* **2023**, *3*, 1104793. [CrossRef]
9. Nagoshi, R.N.; Meagher, R.L.; Hay-Roe, M. Inferring the annual migration patterns of fall armyworm (Lepidoptera: Noctuidae) in the United States from mitochondrial haplotypes. *Ecol. Evol.* **2012**, *2*, 1458–1467. [CrossRef]
10. Pair, S.; Raulston, J.; Rummel, D.; Westbrook, J.; Wolf, W.; Sparks, A.; Schuster, M. Development and production of corn earworm and fall armyworm in the Texas high plains: Evidence for reverse fall migration. *Southwest. Entomol.* **1987**, *12*, 89–99.
11. Mitchell, E.R.; McNeil, J.N.; Westbrook, J.K.; Silvain, J.F.; Lalanne-Cassou, B.; Chalfant, R.B.; Pair, S.D.; Waddill, V.H.; Sotomayor-Rios, A.; Proshold, F.I. Seasonal Periodicity of Fall Armyworm, (Lepidoptera: Noctuidae) in the Caribbean Basin and Northward to Canada. *J. Entomol. Sci.* **1991**, *26*, 39–50. [CrossRef]
12. Gould, F.; Blair, N.; Reid, M.; Rennie, T.; Lopez, J.; Micinski, S. *Bacillus thuringiensis*—Toxin resistance management: Stable isotope assessment of alternate host use by *Helicoverpa zea*. *Proc. Natl. Acad. Sci. USA* **2002**, *99*, 16581–16586. [CrossRef] [PubMed]
13. Westbrook, J.K. Noctuid migration in Texas within the nocturnal aeroecological boundary layer. *Integr. Comp. Biol.* **2008**, *48*, 99–106. [CrossRef] [PubMed]
14. Paula-Moraes, S.V.; Calixto, E.S.; Santos, A.A.; Reay-Jones, F.P.F.; Reisig, D.D.; Farhan, Y.; Smith, J.L.; Hutchison, W.D. Continental-scale migration patterns and origin of *Helicoverpa zea* (Lepidoptera: Noctuidae) based on a biogeochemical marker. *Environ. Entomol.* **2024**, *53*, 487–497. [CrossRef]
15. Walker, T.J. Migrating Lepidoptera: Are butterflies better than moths? *Fla. Entomol.* **1980**, *63*, 79–98. [CrossRef]
16. Stinner, R.; Barfield, C.; Stimac, J.; Dohse, L. Dispersal and movement of insect pests. *Annu. Rev. Entomol.* **1983**, *28*, 319–335. [CrossRef]
17. Hobson, K.A.; Wassenaar, L.I.; Taylor, O.R. Stable isotopes (δD and δ^{13}C) are geographic indicators of natal origins of monarch butterflies in eastern North America. *Oecologia* **1999**, *120*, 397–404. [CrossRef]
18. Wassenaar, L.; Hobson, K. Comparative equilibration and online technique for determination of non-exchangeable hydrogen of keratins for use in animal migration studies. *Isot. Environ. Health Stud.* **2003**, *39*, 211–217. [CrossRef]
19. Ma, C.; Vander Zanden, H.B.; Wunder, M.B.; Bowen, G.J. assignR: An R package for isotope-based geographic assignment. *Methods Ecol. Evol.* **2020**, *11*, 996–1001. [CrossRef]
20. R Core Team. *R: A Language and Environment for Statistical Computing*; R Core Team: Vienna, Austria, 2024.
21. Hobson, K.A.; Kardynal, K.J.; Koehler, G. Expanding the Isotopic Toolbox to Track Monarch Butterfly (*Danaus plexippus*) Origins and Migration: On the Utility of Stable Oxygen Isotope (δ18O) Measurements. *Front. Ecol. Evol.* **2019**, *7*, 224. [CrossRef]
22. Westbrook, J.; Fleischer, S.; Jairam, S.; Meagher, R.; Nagoshi, R. Multigenerational migration of fall armyworm, a pest insect. *Ecosphere* **2019**, *10*, e02919. [CrossRef]
23. Lopez, J.R.; Winter, J.M.; Elliott, J.; Ruane, A.C.; Porter, C.; Hoogenboom, G.; Anderson, M.; Hain, C. Sustainable Use of Groundwater May Dramatically Reduce Irrigated Production of Maize, Soybean, and Wheat. *Earth's Future* **2022**, *10*, e2021EF002018. [CrossRef]
24. Kovač, Z.; Krevh, V.; Filipović, L.; Defterdarović, J.; Balaž, B.-I.; Filipović, V. Estimation of Precipitation Fraction in the Soil Water of the Hillslope Vineyard Using Stable Isotopes of Water. *Water* **2023**, *15*, 988. [CrossRef]
25. Wolf, W.; Westbrook, J.; Raulston, J.; Pair, S.; Hobbs, S. Recent Airborne Radar Observations of Migrant Pests in the United States. *Philos. Trans. R. Soc. Lond. B Biol. Sci.* **1990**, *328*, 619–630. [CrossRef]
26. Beerwinkle, K.R.; Lopez, J.D.; Witz, J.A.; Schleider, P.G.; Eyster, R.S.; Lingren, P.D. Seasonal Radar and Meteorological Observations Associated with Nocturnal Insect Flight at Altitudes to 900 Meters. *Environ. Entomol.* **1994**, *23*, 676–683. [CrossRef]
27. Chapman, J.W.; Drake, V.A.; Reynolds, D.R. Recent Insights from Radar Studies of Insect Flight. *Annu. Rev. Entomol.* **2011**, *56*, 337–356. [CrossRef]
28. Storer, N.P.; Babcock, J.M.; Schlenz, M.; Meade, T.; Thompson, G.D.; Bing, J.W.; Huckaba, R.M. Discovery and Characterization of Field Resistance to Bt Maize: *Spodoptera frugiperda* (Lepidoptera: Noctuidae) in Puerto Rico. *J. Econ. Entomol.* **2010**, *103*, 1031–1038. [CrossRef]

29. Huang, F.; Qureshi, J.A.; Meagher, R.L.; Reisig, D.D.; Head, G.P.; Andow, D.A.; Ni, X.; Kerns, D.; Buntin, G.D.; Niu, Y.; et al. Cry1F Resistance in Fall Armyworm *Spodoptera frugiperda*: Single Gene versus Pyramided Bt Maize. *PLoS ONE* **2014**, *9*, e112958. [CrossRef]
30. Yang, F.; Wang, Z.; Kerns, D.L. Resistance of *Spodoptera frugiperda* to Cry1, Cry2, and Vip3Aa Proteins in Bt Corn and Cotton in the Americas: Implications for the Rest of the World. *J. Econ. Entomol.* **2022**, *115*, 1752–1760. [CrossRef]
31. Gutiérrez-Moreno, R.; Mota-Sanchez, D.; Blanco, C.A.; Whalon, M.E.; Terán-Santofimio, H.; Rodriguez-Maciel, J.C.; DiFonzo, C. Field-Evolved Resistance of the Fall Armyworm (Lepidoptera: Noctuidae) to Synthetic Insecticides in Puerto Rico and Mexico. *J. Econ. Entomol.* **2019**, *112*, 792–802. [CrossRef]
32. Mota-Sanchez, D.; Wise, J. Arthropod Pesticide Resistance Database. Available online: https://www.pesticideresistance.org/ (accessed on 18 November 2024).
33. Dobush, G.R.; Ankney, C.D.; Krementz, D.G. The Effect of Apparatus, Extraction Time, and Solvent Type on Lipid Extractions of Snow Geese. *Can. J. Zool.* **1985**, *63*, 1917–1920. [CrossRef]

insects

MDPI

Article

Dual Role of *Sitophilus zeamais*: A Maize Storage Pest and a Potential Edible Protein Source

Soledad Mora Vásquez * and Silverio García-Lara *

Tecnologico de Monterrey, Escuela de Ingeniería y Ciencias, Ave. Eugenio Garza Sada 2501, Monterrey 64849, Mexico
* Correspondence: sol.mv@tec.mx (S.M.V.); sgarcialara@tec.mx (S.G.-L.)

Simple Summary: Maize experiences significant post-harvest losses due to infestations by *Sitophilus zeamais*. This study investigates the potential of *S. zeamais* as a protein source. The weevils were processed into flour and evaluated for food safety, protein content, and amino acid profile. The resulting flour met safety standards, contained 48.1% protein, and was rich in isoleucine, valine, and threonine, although it lacked some essential amino acids. Despite these limitations, *S. zeamais* flour could serve as a viable protein source for both food and feed applications. Incorporating *S. zeamais* flour into food and feed systems could contribute to improved food security.

Abstract: Maize (*Zea mays*) is a critical staple crop whose post-harvest losses, predominantly due to infestations by the maize weevil, *Sitophilus zeamais*, threaten food security. This study explores the possibility of utilizing *S. zeamais*, traditionally known as a pest, as an alternative protein source by assessing its nutritional profile and food safety attributes. Cultured under controlled conditions, *S. zeamais* specimens were processed into flour, which was subsequently analyzed for microbiological safety, protein content, and amino acid composition. Microbiological assays confirmed that the flour met established food safety standards, with aerobic mesophilic bacteria, fungi, and yeast present at negligible levels and no detection of coliforms, *Salmonella* spp., or *Escherichia coli*. Protein quantification revealed a high total protein content (48.1 ± 0.3%), although the salt-soluble fraction constituted only 13.7% of the total. The amino acid profile exhibited elevated levels of isoleucine, valine, and threonine, while deficiencies in leucine, lysine, sulfur amino acids, and tryptophan were noted. These findings suggest that, despite certain limitations, *S. zeamais* flour represents a viable protein source. Integrating targeted insect harvesting for protein into pest management strategies could help reduce post-harvest losses and contribute to improved food security and nutritional availability.

Keywords: insects; *Sitophilus zeamais*; protein; amino acid profile; food safety; nutritional assessment

1. Introduction

Maize post-harvest losses due to insect infestations pose a major challenge to food security, particularly in maize-dependent regions where maize serves as an important staple crop, providing primary nutrition for human populations and serving as a key component of livestock feed. Exploring alternative uses for insect pests, such as *Sitophilus zeamais*, could help mitigate these losses while contributing to sustainable protein sources for both food and feed applications. Among the most destructive stored-product pests is *Sitophilus zeamais* Motschulsky (Coleoptera: Curculionidae), commonly known as the maize weevil—an insect species that significantly reduces grain weight, depletes nutritional quality, and impairs germination potential, leading to substantial economic losses [1]. *S. zeamais* thrives in warm and humid environments, with females capable of laying up to 575 eggs within maize kernels, where the larvae develop internally, consuming the endosperm and reducing grain integrity. Infestations are widespread across tropical and subtropical regions, including Asia, Africa, and the Americas, where post-harvest storage systems remain vulnerable to weevil proliferation [2].

Traditional pest management strategies for *S. zeamais* include chemical fumigation, hermetic storage, and biological control methods, such as the use of parasitoid wasps (*Anisopteromalus calandrae*) and entomopathogenic fungi (*Beauveria bassiana*) [3,4]. The growing challenges associated with insecticide resistance, environmental degradation, and pesticide residue accumulation have intensified the search for alternative strategies within integrated pest management frameworks [5]. Given the increasing interest in alternative and sustainable protein sources [6], researchers are increasingly exploring new edible insect species [7] that may offer innovative opportunities to address both food system vulnerabilities and pest control. In this context, the potential utilization of *Sitophilus zeamais* as an edible insect remains largely unexplored. This approach could provide dual benefits of reducing post-harvest maize losses through targeted pest harvesting while also contributing to food security by supplying a sustainable source of protein and essential nutrients.

Insect consumption (entomophagy) is a widespread practice in many cultures, with more than 1900 insect species recognized as edible [8]. Various stored-product pests have historically been consumed, including locusts (*Schistocerca gregaria*), palm weevils (*Rhynchophorus* spp.), and termites (*Macrotermes bellicosus*), demonstrating that the harvesting of pest species for animal or human consumption can serve as a sustainable food production strategy while reducing agricultural losses [9–12]. In the case of *S. zeamais*, previous studies have documented its consumption in Ghana [13], Nigeria [14], and the Philippines [15]. These reports highlight its chemical composition and mineral content, yet its full nutritional profile, particularly its protein and amino acid composition, remains underexplored.

This study aims to evaluate the nutritional profile and edibility of *Sitophilus zeamais*, focusing on its potential as a sustainable protein source while considering its implications for pest control strategies in maize storage systems. By assessing its protein content, amino acid composition, and microbiological safety, this research contributes to the broader discourse on stored-product pest management, maize post-harvest preservation, and alternative protein sources. Recognizing the emerging potential of *S. zeamais* as a food source could create new opportunities for integrating sustainable pest management strategies while contributing to food security solutions.

2. Materials and Methods

2.1. Insect Pest Culture

The cultivation of *Sitophilus zeamais* was carried out in the Postharvest Biotechnology Laboratory at Tecnológico de Monterrey, Mexico. Adult specimens of *S. zeamais* were collected from stored maize in Agua Fría, Mexico, and cultured on white maize (single-cross dent hybrid) for four generational cycles under controlled conditions: 27 ± 1 °C, 70 ± 5% relative humidity (RH), and a 12:12 h light/dark photoperiod [16]. Insects were reared in 16 oz (473 mL) Regular Mouth Mason jars (approximately 12.7 cm in height and 7.6 cm in diameter) with airtight lids and bands (Ball®, Newell Brands Inc., Atlanta, GA, USA). After two months of cultivation, adult insects were collected, washed with distilled water, surface-disinfected using 90% ethanol for 2 min, and subsequently dried. The dried insects were ground into a fine powder using a cyclone mill equipped with a 1 mm screen to ensure uniformity of the sample. The resulting powdered insect material was used in subsequent biochemical and microbiological analyses.

2.2. Food Safety Analysis

A microbiological assay for food safety was carried out following the guidelines stated by the standard methods (Mexican Official Norms NOM-122-SSA1-1994 [17]). For the analysis, 10 g of insect powder was weighed into sterile containers and diluted with 90 mL of sterile diluent. Prior to analysis, the frozen sample was thawed under refrigeration (4–8 °C) for 18 to 24 h, following standard microbiological preparation procedures [18]. Samples were analyzed for aerobic mesophilic bacteria, fungi, yeast, total coliforms, *Escherichia coli*, *Staphylococcus aureus*, and *Salmonella* spp. Aerobic mesophilic bacteria were cultured on Plate Count Agar [18], fungi and yeast on Potato Dextrose Agar [19], and total coliforms and *E. coli* on Brilliant Green Bile Broth [20]. *Staphylococcus aureus* was identified using Baird-Parker medium [21], and *Salmonella* spp. on Xylose Lysine Deoxycholate (XLD) or Hektoen Enteric Agar [22]. All culture media and microbiological reagents were purchased from Sigma-Aldrich (St. Louis, MO, USA), unless otherwise stated. All inoculated plates were incubated under conditions based on standard procedures, including temperature, time, and atmosphere appropriate to the target microorganism [17]. After incubation, microbial colonies were enumerated and key morphological traits—such as size, shape, color, and texture—were assessed in accordance with standard microbiological protocols. All analyses were performed in triplicate.

2.3. Extraction of Salt-Soluble Proteins

Salt-soluble proteins were extracted from the sample following the method described by Kim et al. [23]. Specifically, 9 mL of a 0.5 M saline solution was added to 1.5 g of finely ground insect sample in a test tube containing 3 g of glass beads. The test tube was then placed in a shaker and incubated at 4 °C for 2 h. After incubation, the sample was centrifuged at 10,000× *g* for 20 min. A second extraction was performed on the resulting pellet by adding an additional 9 mL of saline solution, shaking the mixture at 4 °C for 1 h, and centrifuging under the same conditions. The supernatants obtained from both extraction steps were combined to yield the salt-soluble protein fraction of the insect sample [23]. Our experiments, conducted in triplicate, confirmed that water-soluble proteins were negligible, thereby justifying the exclusive focus on salt-soluble proteins.

2.4. Protein Quantification

Total and salt-soluble protein content was determined using the Kjeldahl method (AOAC Method 928.08), applying a nitrogen-to-protein conversion factor of 5.30 instead of the conventional 6.25. This adjustment was made to mitigate protein overestimation, as the insect cuticle contains substantial amounts of fibrous chitin along with proteins that are tightly embedded within its matrix [24]. A total of 0.1 g of the sample was placed into a digestion flask containing 0.05 g of $CuSO_4$ and 1.95 g of K_2SO_4, and subsequently 3 mL of H_2SO_4 was added. The mixture was digested on a heating grill for 1 h. After digestion, the mixture was diluted with 10 mL of distilled water, followed by the addition of 10 mL of 50% NaOH. The resulting solution was then distilled into a receiver containing an indicator solution, and titration was carried out with 0.200 N HCl until the sample turned transparent [25].

2.5. Protein Quality

Approximately 500 mg of *Sitophilus zeamais* flour was accurately weighed and subjected to hydrolysis. For the stable amino acids, isoleucine, leucine, lysine, phenylalanine, threonine, valine, and tyrosine, samples (approximately 500 mg) were hydrolyzed in 6 N

HCl containing 0.1% phenol at 110 °C for 24 h under a nitrogen atmosphere. In contrast, tryptophan, which is labile under acidic conditions, was hydrolyzed using 4 M NaOH at 110 °C for 16–18 h, followed by neutralization prior to analysis. After hydrolysis, the sample was cooled to room temperature and the hydrolysate was filtered through a 0.45 µm membrane filter. The filtrate was subsequently evaporated to dryness under reduced pressure at 40 °C and reconstituted in 5 mL of mobile phase. The reconstituted sample was analyzed using high-performance liquid chromatography (HPLC) coupled with an evaporative light scattering detector (ELSD; Agilent Technologies, Santa Clara, CA, USA). Separation was achieved on a reversed-phase C18 column (4.6 mm × 250 mm, 5 µm particle size) maintained at 40 °C. The mobile phases consisted of (A) 0.1% trifluoroacetic acid (TFA) in water and (B) acetonitrile. A gradient elution program was employed, starting at 95% A and 5% B, with a gradual increase in the proportion of B over a 30 min run time to achieve optimal separation of individual amino acids. The flow rate was maintained at 1.0 mL/min, and the injection volume was set to 20 µL. The ELSD was operated under the following conditions: nebulizer gas (nitrogen) flow rate was set to 2.5 L/min, the drift tube temperature was maintained at 90 °C, and the detector gain was optimized to ensure maximum sensitivity for the analytes of interest. Calibration curves were generated using standard solutions of individual amino acids prepared in the same mobile phase, covering a range of concentrations to ensure accurate quantification [26]. The experiment was performed in duplicate. The obtained results were subsequently analyzed by comparing them with the amino acid requirements for infants during the growth stage, as specified by the Food and Agriculture Organization [27].

2.6. Statistical Analysis

All primary parameters were expressed as means ± standard deviations. Statistical analyses were conducted using analysis of variance (ANOVA) in the Minitab 19 statistical software (Minitab Inc., State College, PA, USA).

3. Results

3.1. Food Safety Analysis

The microbiological analysis of Sitophilus zeamais flour (Table 1) showed aerobic mesophilic bacteria at 590 UFC/g, well below the maximum limit of 100,000 UFC/g. Fungi, yeast, and total coliforms were not detected (<10 UFC/g), and Staphylococcus aureus was present at <10 UFC/g, within the allowable limit of 100 UFC/g. *Salmonella* spp. and Escherichia coli were absent.

Table 1. Microbiological test for food safety analysis of *Sitophilus zeamais* flour.

Microorganism	Result (UFC/g) **	Maximum Limit (UFC/g) *
Aerobic mesophilic bacteria	590	100,000
Fungi	<10	<10
Yeast	<10	<10
Total coliform	None	100
Staphylococcus aureus	<10	100
Salmonella spp.	None	Negative
Escherichia coli	None	Negative

* In accordance with the Health Secretary of Mexico in order to guarantee the quality of food safety analysis (NOM-122-SSA1-1994). ** Data represent the mean of three independent replicates.

3.2. Protein Quantification

The protein content of *Sitophilus zeamais* flour was 48.1 ± 0.3% on a dry matter basis, with a salt-soluble protein fraction of 6.6 ± 1.3%, representing approximately 13.7% of the total protein (Table 2).

Table 2. Total protein and protein profile of *Sitophilus zeamais* flour.

Protein	Content (%) *
Salt-soluble fraction	6.6 ± 1.3
Total Content	48.1 ± 0.3

* Mean values ± standard deviation (n = 3).

3.3. Protein Quality and Amino Acid Composition

The amino acid profile of Sitophilus zeamais raw flour (Table 3) shows that isoleucine, valine, and threonine exceeded FAO reference values by 40%, 65%, and 24%, respectively. Aromatic amino acids (phenylalanine and tyrosine) were present at more than double the recommended concentration, with an amino acid score of 2.0. In contrast, leucine, lysine, sulfur amino acids (methionine and cysteine), and tryptophan had lower amino acid scores, ranging from 0.7 to 0.9.

Table 3. Amino acid profile and protein quality of *Sitophilus zeamais* raw flour.

Amino Acid	Reference (mg/g Protein) *	*S. zeamais* (mg/g Protein) **	Difference	Amino Acid Score ***
Isoleucine	30	42	+12	1.4
Leucine	61	55.4	−5.6	0.9
Lysine	48	40.6	−7.4	0.8
Methionine	-	10.9	-	
Cysteine	-	5.6	-	
Sulfur AA (Met + Cys)	23	16.5	−6.5	0.7
Phenylalanine	-	30.5	-	
Tyrosine	-	53.3	-	
Aromatic AA (Phe + Tyr)	41	83.8	+43	2.0
Tryptophan	6.6	4.8	−1.8	0.7
Valine	40	66.3	+24	1.7
Threonine	25	30.9	+5.9	1.2

* Reference based on recommended AA (amino acids) for an infant at growing stage [27]. ** Data correspond to the average of two determinations with an analytical variability of less than 1%. *** *Amino acid score* = $\frac{mgAA \text{ in } 1\,g \text{ of } S.zeamais \text{ protein}}{mg\ AA \text{ in } 1\,g \text{ of reference protein}}$.

4. Discussion

4.1. Food Safety Analysis

The microbiological analysis of *Sitophilus zeamais* flour indicates that the product complies with established food safety standards. Aerobic mesophilic bacteria were present at 590 UFC/g, a value significantly below the maximum allowable limit of 100,000 UFC/g as defined by NOM-122-SSA1-1994 [17]. Moreover, the levels of fungi and yeast were below the detection threshold (<10 UFC/g), and total coliform bacteria were not detected. *Staphylococcus aureus* was identified at <10 UFC/g, which is well within the permissible limit of 100 UFC/g. Notably, *Salmonella* spp. and *Escherichia coli* were absent from the sample, further substantiating the microbiological safety of the flour.

Previous research has demonstrated that microbial loads in edible insect products are influenced by multiple factors, including the insects' inherent microbial content, the impact of processing on bacterial populations, and the risk of secondary contamination [28]. When subjected to appropriate processing techniques, insect-derived flours and other products exhibit microbial profiles that meet or exceed food safety standards. These consistent outcomes across various studies suggest that the application of standardized hygienic practices and controlled processing conditions is crucial for ensuring the safety of edible insect products [29].

From a pest control perspective, these findings are relevant because they indicate that harvesting *S. zeamais* from maize storage facilities for consumption does not introduce additional food safety concerns. If integrated into post-harvest pest management strategies, targeted collection efforts could help reduce infestation rates in stored grains, providing an alternative pest mitigation approach while ensuring nutritional benefits [30].

4.2. Protein Quantification

The high total protein content confirms that *S. zeamais* flour is a protein-rich material, a characteristic that is frequently reported in studies focusing on edible insects. However, the salt-soluble fraction, which constitutes approximately 13.7% of the total protein content, represents only a minor component of the overall protein profile. This observation is significant, as salt-soluble proteins generally include those involved in enzymatic functions and other cellular activities that require ionic interactions for stability and solubility. In contrast, the bulk of the protein content may be composed of proteins with different solubility properties, such as water-insoluble or structural proteins.

The relatively low proportion of salt-soluble proteins could be attributed to the biological characteristics and functional roles of proteins within *S. zeamais*. It is conceivable that the majority of the proteins are either bound to cellular structures or exist in forms that do not readily solubilize in saline solutions. In studies involving *Tenebrio molitor*, the salt-soluble protein fraction has been shown to be significantly more digestible compared to the insoluble fraction. This soluble fraction was notably enriched in hemolymph proteins and enzymes such as alpha-amylase, which play essential roles in nutrient transport and carbohydrate metabolism [31]. These findings suggest that the high digestibility of the soluble proteins is largely attributable to their specific composition, which favors proteins involved in physiological functions over more structurally bound muscle proteins. This distinction in protein solubility has implications for both the biological understanding of the insect and potential control strategies. For instance, detailed characterization of the protein profile could provide insights into metabolic pathways critical for insect survival, thereby identifying novel targets for pest management interventions.

Moreover, the substantial overall protein content underscores the potential for utilizing *S. zeamais* as a source of protein in various applications, including animal feed and human food products, provided that safety and processing standards are met. The identification of specific protein fractions, such as the salt-soluble fraction, may also aid in the development of extraction techniques that maximize yield and functional quality, ultimately contributing to the valorization of insect biomass in sustainable food systems.

Numerous studies have quantified the protein content in various insect families, reporting high overall values: Saturniidae (40–50%), Notodontidae (42–45%), Gryllidae (53%), Acrididae (76%), and Tenebrionidae (52%) [31–34]. *Sitophilus zeamais* flour exhibits a total protein content of 48.1 ± 0.3%. This value falls within the range observed for other insect species, underscoring the potential of *S. zeamais* as a protein-rich resource for nutritional applications.

In conclusion, the protein profile of *S. zeamais* flour, characterized by a high total protein content and a modest salt-soluble fraction, offers valuable insights into the insect's biology. These findings not only enhance our understanding of protein composition in corn insect pests but also pave the way for future research aimed at exploiting these biological resources for innovative control and utilization technologies.

4.3. Protein Quality and Amino Acid Composition

The amino acid profile of *Sitophilus zeamais* raw flour, as detailed in Table 3, reveals a complex pattern of nutritional adequacy and limitation when compared to the FAO-recommended amino acid requirements for infants in the growing stage [27]. Notably, the levels of isoleucine, valine, and threonine exceed the reference values by 40%, 65%, and 24%, respectively, and the aromatic amino acids (combined phenylalanine and tyrosine) are present at more than double the recommended concentration, yielding an amino acid score of 2.0. Conversely, leucine, lysine, sulfur amino acids (methionine and cysteine combined), and tryptophan fall below the reference levels, with amino acid scores ranging from 0.7 to 0.9, thereby identifying them as limiting factors in the protein quality of the flour. This duality in amino acid composition suggests that while *S. zeamais* flour could serve as a valuable protein source, particularly in applications requiring high levels of certain essential amino acids, its deficiencies in others may necessitate formulation adjustments or supplementation to achieve a balanced nutritional profile.

These results align with previous studies on edible insects, which have similarly reported high protein content with favorable levels of certain essential amino acids, though often with one or more limiting amino acids that restrict the flour's use as a complete protein source [33]. Therefore, while *S. zeamais* raw flour exhibits promising nutritional attributes, its integration into food formulations may require strategies to overcome its amino acid limitations. These could include fortification with sulfur-containing amino acids, such as cysteine or methionine, or blending with complementary protein sources—for example, legumes (e.g., soy or lentils), eggs, or dairy proteins—that provide the deficient amino acids [35]. Such approaches are commonly used in food product development to improve amino acid balance and enhance the overall nutritional quality of novel protein ingredients.

4.4. Pest Control and Sustainable Utilization Strategies

Several precedents have demonstrated the potential of utilizing agricultural pests as food sources to mitigate their negative impacts. For instance, locust harvesting—specifically of *Schistocerca gregaria* and *Locusta migratoria*—has contributed to reducing swarm sizes in Africa while simultaneously providing a high-protein food source [36]. Similarly, the collection of palm weevil larvae (*Rhynchophorus* spp.) has been shown to lower infestation rates in oil palm plantations, with these larvae consumed as a delicacy in West Africa and Southeast Asia [37]. Additionally, grasshopper collection in Uganda has proven effective in managing outbreaks of *Ruspolia differens* [38]. These examples suggest that harvesting pests for consumption can serve as a viable control strategy. In light of these findings, further research should evaluate the feasibility of implementing controlled *Sitophilus zeamais* collection programs. Such programs could reduce post-harvest losses in maize storage facilities, provide a locally available protein source in regions facing food insecurity, and diminish pesticide dependency by incorporating entomophagy into pest management strategies. Moreover, assessing consumer acceptance and examining regulatory frameworks will be essential to ensure the scalability and economic viability of this integrated approach.

5. Conclusions

This study highlights the dual role of *Sitophilus zeamais* as both an agricultural pest and a potential alternative protein source. The high protein content and essential amino acid composition of *S. zeamais* support its nutritional viability, while its presence in maize storage systems suggests an opportunity for sustainable pest control through targeted harvesting.

By integrating biological control methods with entomophagy, it may be possible to mitigate post-harvest losses while contributing to food security and sustainable agriculture.

While the present study provides an initial characterization of *S. zeamais* from a nutritional and microbiological perspective, it should be regarded as a preliminary step toward broader evaluations. A comprehensive assessment of its safety for human consumption—including allergenic potential, toxicological risks, and long-term effects—remains essential. Accordingly, future research should address these dimensions, in addition to exploring scalability, consumer acceptance, and regulatory frameworks, to determine the feasibility of incorporating *S. zeamais* into integrated pest management and sustainable food systems.

Author Contributions: Conceptualization S.G.-L.; methodology, S.G.-L.; validation, S.G.-L.; analysis, S.M.V. and S.G.-L.; writing—original draft preparation, S.M.V. and S.G.-L.; writing—review and editing, S.M.V. and S.G.-L. All authors have read and agreed to the published version of the manuscript.

Funding: This research was supported by the Research Chair Funds CAT-005 from Tecnologico de Monterrey-Campus Monterrey.

Data Availability Statement: The data presented in this study are available on request from the corresponding author due to confidentiality agreements with collaborating institutions. Graphic Abstract created in BioRender, https://BioRender.com/gexem4m (accessed on 6 March 2025).

Acknowledgments: The authors are thankful to Scarlett Puebla, Amparo Duran, and Sofia Ortiz, for technical and scientific help and support on this study. We would like to respectfully acknowledge the memory of Sergio Serna Saldívar, whose contributions and support to this work were invaluable.

Conflicts of Interest: The authors declare no conflicts of interest.

References

1. García-Lara, S.; Saldivar, S.O.S. Insect Pests. In *Encyclopedia of Food and Health*; Elsevier: Amsterdam, The Netherlands, 2016; pp. 432–436.
2. Arrahman, A.; Mirsam, H.; Djaenuddin, N.; Suriani; Pakki, S.; Saenong, M.S.; Sebayang, A. An In-Depth Study on *Sitophilus zeamais* Motsch (Coleoptera: Curculionidae) Pests on Corn Plants. *IOP Conf. Ser. Earth Env. Sci.* **2022**, *1107*, 012060. [CrossRef]
3. Teshome, A.; Tefera, T. Susceptibility of *Sitophilus zeamais* (Mostch.) (Coleoptera: Curculionidae) to *Beauveria bassiana* and *Metarhizium anisopliae*. *SINET Ethiop. J. Sci.* **2009**, *32*, 21–28. [CrossRef]
4. Arbogast, R.T.; Mullen, M.A. Interaction of *Maize weevil* (Coleoptera: Curculionidae) and Parasitoid *Anisopteromalus calandrae* (Hymenoptera: Pteromalidae) in a Small Bulk of Stored Corn. *J. Econ. Entomol.* **1990**, *83*, 2462–2468. [CrossRef]
5. Cortese, D.; Oliveira, G.S.; de Fernandes, M.G. Resistance Evaluation of Maize Varieties to *Sitophilus zeamais* Infestation across Two Generations: Insights for Integrated Pest Management. *J. Stored Prod. Res.* **2024**, *109*, 102473. [CrossRef]
6. Liceaga, A.M.; Aguilar-Toalá, J.E.; Vallejo-Cordoba, B.; González-Córdova, A.F.; Hernández-Mendoza, A. Insects as an Alternative Protein Source. *Annu. Rev. Food Sci. Technol.* **2022**, *13*, 19–34. [CrossRef] [PubMed]
7. Li, M.; Mao, C.; Li, X.; Jiang, L.; Zhang, W.; Li, M.; Liu, H.; Fang, Y.; Liu, S.; Yang, G.; et al. Edible Insects: A New Sustainable Nutritional Resource Worth Promoting. *Foods* **2023**, *12*, 4073. [CrossRef] [PubMed]
8. FAO. 'Worm' Up to the Idea of Edible Insects. Available online: https://www.fao.org/newsroom/story/-Worm-up-to-the-idea-of-edible-insects/en (accessed on 24 January 2025).
9. Payne, C.; Van Itterbeeck, J. Ecosystem Services from Edible Insects in Agricultural Systems: A Review. *Insects* **2017**, *8*, 24. [CrossRef]
10. Adepoju, O.T.; Ajayi, K. Assessment of Quality and Safety of Winged Termites (*Macrotermes bellicosus*) Enriched Locally Formulated Complementary Foods. *J. Food Res.* **2017**, *6*, 117. [CrossRef]
11. Siddiqui, S.A.; Thanpandiyan, K.; Adli, D.N.; Yudhistira, B.; Fernando, I.; De Palo, P. Overview of the African Palm Weevil (*Rhynchophorus phoenicis*) as Food and Feed—A Critical Review. *J. Insects Food Feed.* **2024**, *11*, 593–620. [CrossRef]
12. Abd-El Wahed, S.; Ahmad, A. Variations in Chemical Composition Value of Adults and Nymphs Desert Locust, *Schistocerca gregaria* Forskal (Orthoptera: Acrididae). *J. Plant Prot. Pathol.* **2019**, *10*, 677–681. [CrossRef]
13. Aydoğan, Z.; Gürol, A.; İncekara, Ü.; Tahidu, O.D. Element Content Analysis of Edible Insect of Ghana (Curculionidae: *Sitophilus zeamais*) Using EDXRF Spectrometer. *Erzincan Üniversitesi Fen Bilim. Enstitüsü Derg.* **2016**, *9*, 86–94. [CrossRef]

14. Adeyeye, E.I.; Olaleye, A.A. Chemical Composition and Mineral Safety Index of Five Insects Commonly Eaten in South West Nigeria. *FUW Trends Sci. Technol. J.* **2016**, *1*, 139–144.
15. Feresin, C.; Močinić, S.; Tatković, N. Is Insect Eating Festival a Viable Strategy in Promoting Entomophagy? An Investigation in Marawi City, Philippines. *Int. J. Humanit. Soc. Sci.* **2018**, *10*, 72–78.
16. García-Lara, S.; Khairallah, M.M.; Vargas, M.; Bergvinson, D.J. Mapping of QTL Associated with Maize Weevil Resistance in Tropical Maize. *Crop Sci.* **2009**, *49*, 139–149. [CrossRef]
17. *NOM-122-SSA1-1994*; Productos de La Carne. Productos Cárnicos Curados y Cocidos, y Curados Emulsionados y Cocidos. Especificaciones Sanitarias. Secretaria de Salud: Acapulco, Mexico, 1994. Available online: https://salud.gob.mx/unidades/cdi/nom/122ssa14.html (accessed on 1 February 2025).
18. *NOM-092-SSA1-1994*; Método Para La Cuenta de Bacterias Aerobias En Placa. Sanitarias. Secretaria de Salud: Acapulco, Mexico, 1994. Available online: https://www.dof.gob.mx/nota_detalle.php?codigo=4886029&fecha=12/12/1995#gsc.tab=0 (accessed on 1 February 2025).
19. *NOM-111-SSA1-1994*; Método Para La Cuenta de Mohos y Levaduras En Alimentos. Sanitarias. Secretaria de Salud: Acapulco, Mexico, 1994. Available online: https://dof.gob.mx/nota_detalle.php?codigo=4881226&fecha=13/09/1995#gsc.tab=0 (accessed on 1 February 2025).
20. *NOM-112-SSA1-1994*; Determinación de Bacterias Coliformes. Sanitarias. Secretaria de Salud: Acapulco, Mexico, 1994. Available online: https://dof.gob.mx/nota_detalle.php?codigo=4728925&fecha=15/08/1994#gsc.tab=0 (accessed on 1 February 2025).
21. *NOM-115-SSA1-1994*; Método Para La Determinación de Staphylococcus Aureus En Alimentos. Sanitarias. Secretaria de Salud: Acapulco, Mexico, 1994. Available online: https://www.dof.gob.mx/nota_detalle.php?codigo=4869580&fecha=20/02/1995#gsc.tab=0 (accessed on 1 February 2025).
22. *NOM-114-SSA1-1994*; Método Para La Determinación de Salmonella En Alimentos. Sanitarias. Secretaria de Salud: Acapulco, Mexico, 1994. Available online: https://www.dof.gob.mx/nota_detalle.php?codigo=4881851&fecha=22/09/1995#gsc.tab=0 (accessed on 1 February 2025).
23. Kim, T.-K.; Yong, H.I.; Jeong, C.H.; Han, S.G.; Kim, Y.-B.; Paik, H.-D.; Choi, Y.-S. Technical Functional Properties of Water- and Salt-Soluble Proteins Extracted from Edible Insects. *Food Sci. Anim. Resour.* **2019**, *39*, 643–654. [CrossRef] [PubMed]
24. Boulos, S.; Tännler, A.; Nyström, L. Nitrogen-to-Protein Conversion Factors for Edible Insects on the Swiss Market: *T. Molitor*, *A. Domesticus*, and *L. Migratoria*. *Front. Nutr.* **2020**, *7*, 89. [CrossRef]
25. Sáez-Plaza, P.; Michałowski, T.; Navas, M.J.; Asuero, A.G.; Wybraniec, S. An Overview of the Kjeldahl Method of Nitrogen Determination. Part I. Early History, Chemistry of the Procedure, and Titrimetric Finish. *Crit. Rev. Anal. Chem.* **2013**, *43*, 178–223. [CrossRef]
26. Eakkanaluksamee, K.; Anuntagool, J. Optimization of High-Protein Glutinous Rice Flour Production Using Response Surface Method. *Rice Sci.* **2020**, *27*, 75–80. [CrossRef]
27. FAO. Dietary Protein Quality Evaluation in Human Nutrition Report of an FAO Expert Consultation. In *FAO Food and Nutrition Paper*; FAO: Rome, Italy, 2013; Volume 92.
28. Grabowski, N.T.; Klein, G. Microbiology of Processed Edible Insect Products—Results of a Preliminary Survey. *Int. J. Food Microbiol.* **2017**, *243*, 103–107. [CrossRef]
29. Lange, K.W.; Nakamura, Y. Edible Insects as Future Food: Chances and Challenges. *J. Future Foods* **2021**, *1*, 38–46. [CrossRef]
30. Cohnstaedt, L.W.; Lado, P.; Ewing, R.; Cherico, J.; Brabec, D.; Shults, P.; Arsi, K.; Donoghue, A.M.; Wagner, R.; Chaskopoulou, A. Harvesting Insect Pests for Animal Feed: Potential to Capture an Unexploited Resource. *J. Econ. Entomol.* **2024**, *117*, 1301–1305. [CrossRef] [PubMed]
31. Yi, L.; Van Boekel, M.A.J.S.; Boeren, S.; Lakemond, C.M.M. Protein Identification and in Vitro Digestion of Fractions from Tenebrio Molitor. *Eur. Food Res. Technol.* **2016**, *242*, 1285–1297. [CrossRef]
32. Payne, C.L.R.; Scarborough, P.; Rayner, M.; Nonaka, K. A Systematic Review of Nutrient Composition Data Available for Twelve Commercially Available Edible Insects, and Comparison with Reference Values. *Trends Food Sci. Technol.* **2016**, *47*, 69–77. [CrossRef]
33. Zielińska, E.; Baraniak, B.; Karaś, M.; Rybczyńska, K.; Jakubczyk, A. Selected Species of Edible Insects as a Source of Nutrient Composition. *Food Res. Int.* **2015**, *77*, 460–466. [CrossRef]
34. Nsevolo Miankeba, P.; Taofic, A.; Kiatoko, N.; Mutiaka, K.; Francis, F.; Caparros Megido, R. Protein Content and Amino Acid Profiles of Selected Edible Insect Species from the Democratic Republic of Congo Relevant for Transboundary Trade across Africa. *Insects* **2022**, *13*, 994. [CrossRef]
35. Dimina, L.; Rémond, D.; Huneau, J.-F.; Mariotti, F. Combining Plant Proteins to Achieve Amino Acid Profiles Adapted to Various Nutritional Objectives—An Exploratory Analysis Using Linear Programming. *Front. Nutr.* **2022**, *8*, 809685. [CrossRef]
36. van Huis, A. Harvesting Desert Locusts for Food and Feed May Contribute to Crop Protection but Will Not Suppress Upsurges and Plagues. *J. Insects Food Feed.* **2021**, *7*, 245–248. [CrossRef]

37. El-Shafie, H.A.F. Utilization of Edible Insects as Food and Feed with Emphasis on the Red Palm Weevil. In *Food and Nutrition Security in the Kingdom of Saudi Arabia, Vol. 2: Macroeconomic Policy and Its Implication on Food and Nutrition Security*; Springer: Cham, Switzerland, 2024.
38. John, B.; Lawrence, T.; Francis, S.; Alfonce, L.; Peter, E.J. Opportunities, Challenges and Policy Gaps for Sustainable Consumption of Edible Long-Horned Grasshopper, Ruspolia Differens, in Uganda: A Review. *Int. J. Trop. Insect Sci.* **2024**, *44*, 2665–2675. [CrossRef]

insects

MDPI

Article

Differential Characterization of Midgut Microbiota Between Bt-Resistant and Bt-Susceptible Populations of *Ostrinia furnacalis*

Juntao Zhang, Ziwen Zhou, Xiaobei Liu, Yongjun Zhang and Tiantao Zhang *

State Key Laboratory for Biology of Plant Diseases and Insect Pests, Institute of Plant Protection, Chinese Academy of Agricultural Sciences, Beijing 100193, China

* Correspondence: zhangtiantao@caas.cn

Simple Summary: *Bacillus thuringiensis* (Bt) is extensively used all over the world as a type of eco-friendly pesticide. Despite its high efficacy against various pests, the main challenge is the emergence of pest resistance. In this study, we verified the differences of midgut microbiota in four Bt-resistant strains and certified that *Enterococcus* enhances the Cry1Ab resistance of the Asian corn borer by bioassays.

Abstract: *Bacillus thuringiensis* (Bt) is an efficacious biocontrol bacterium known for producing various toxins, such as crystal toxins, which disrupt the midgut epithelium of pest larvae, leading to larval mortality. However, the development of resistance to *Bacillus thuringiensis* in pests poses a significant threat to the widespread application of Bt corn. Consequently, we employed high-throughput sequencing of the midgut bacterial 16S ribosomal RNA to characterize the midgut bacteria in four Bt-resistant strains. Specifically, Bt-resistant strains (ACB-FR and ACB-AcR) exhibited lower bacterial diversity compared to ACB-AbR and ACB-IeR. Multivariate analyses and statistical evaluations further demonstrated that the microbiota communities in Bt-resistant pests (AbR, AcR, IeR, and FR) were distinct from those in Bt-susceptible strains. Notably, the genus *Klebsiella* predominated in BtS, whereas *Enterococcus* was the genus with peak enrichment in AbR, AcR, IeR, and FR. Bioassays subsequently revealed that *Enterococcus* enhances the Cry1Ab resistance of ACB larvae. Our investigations indicate that treatment with Bt protein alters the midgut microbiota community of *O. furnacalis*, and these microbiota differences may potentially modulate the Bt-induced lethality mechanism.

Keywords: *Ostrinia furnacalis*; midgut bacteria; *Bacillus thuringiensis*; Cry toxin

1. Introduction

Bacillus thuringiensis (Bt), is a ubiquitous Gram-positive, endospore-forming bacterium originally characterized as an entomopathogen due to exhibiting insecticidal activity primarily through crystalline (Cry) proteins produced during sporulation, which accumulate as parasporal inclusions [1]. These production are primarily composed of one or more proteins known as crystal (Cry) and cytolitic (Cyt) toxins, also referred to as δ-endotoxins [2]. These proteins (protoxins) are activated by insecticidal midgut proteases, and the activated toxins affect the larval midgut epithelium, causing a collapse of the membrane and ultimately leading to insect death [3]. Bt gene-modified crops have been globally adopted as a sustainable for agricultural pest control and human disease vector management. Nevertheless, the emergence of resistance in pests compromises the sustainable deployment of Bt-based crop protection strategies [4].

The Asian corn borer, *Ostrinia furnacalis* (Guenée), which is abbreviated as ACB, is one of the most economically significant pests of maize in China. Research on the Asian corn borer has been conducted since the 1930s [5]. The larvae exhibit broad feeding behavior across all maize tissues, of which the most severe crop damage results from their drilling activity in developing ears and stalks [6]. The Asian corn borer can cause significant maize losses, estimated at 6–9 million tons for an ordinary year. However, most maize fields are left without any effective resistance measures. Field tests in China have shown that maize expressing the Bt toxin (MON810) have significantly increased ACB larval mortality [7]. However, quantitative bioassays revealed that lab-selected ACB strains achieve resistance ratios (RR_{50}) of >150 (Cry1Ab), 350 (Cry1Ac), >800 (Cry1Ie), and 1700 (Cry1F) [8–11]. This rapid evolution of resistance may severely limit the expanding application of genetically modified maize.

Laboratory and field data have identified three different resistant mechanisms in pests to Bt crops, namely, altered toxin activation, mutations in toxin-binding receptors, and the immune response system [12]. Comparing the toxin receptor sequences between Cry1Ab-resistant and -susceptible ACB colonies, the resistant colony decreased some or all of their binding capacity, presumably by altering one or more of its shared binding sites [13]. Research and laboratory experiments demonstrated that Cry1 proteins bind to several membrane-associated receptors in the midgut epithelium of *Lepidopteran* larvae, such as aminopeptidase N, cadherin, and ABC transporters [14–17]. Several studies provide evidence confirming that the r1–r3 alleles of the Lepidopterous larvae cadherin gene BtR confer recessively inherited resistance to the Cry1Ac toxin [18]. Nevertheless, the accurate mechanism underlying ACB resistance to Bt remains not fully proven. The insect gut contains many different kinds of microbiota that may contribute to insect hosts surviving and adapting to the environment [19–21]. Numerous studies have revealed that midgut bacteria can influence the pest resistance of Bt while potentially contributing to the insect Bt resistance development [22]. Mortality caused by Cry1 toxin exposure in species such as *Vanessa cardui, Manduca sexta,* and *Pieris rapae* was reduced under an antibiotics diet to decrease the gut microbiota. However, reintroducing the native *Enterobacter* sp. restored high toxicity [23]. Bt toxin susceptibility was also found to decrease in *Plodia interpunctella* after removing the gut bacteria [24]. These experiments demonstrate that gut bacteria critically modulate Bt toxin efficacy to pests, especially in Lepidoptera.

To investigate potential associations between midgut microbiota and Bt resistance in ACB, we compared the microbial communities of five ACB strains (BtS, AbR, AcR, IeR, and FR) using 16S rRNA Illumina sequencing. Our study investigated how gut microbiota modulate Bt resistance and susceptibility in the Asian corn borer with a focus on characterizing functional differences and interactions between resistant and susceptible host strains.

2. Materials and Methods

2.1. Mass-Rearing and Artifical Selection of Bt-Resistant ACB Strains

Five laboratory strains of ACB were used in this research, namely, a Bt-susceptible strain (S) and four laboratory resistant strains (BtR) selected under different Cry toxins: Cry1Ab (AbR), Cry1Ac (AcR), Cry1F (FR), and Cry1Ie (IeR).

All colonies were obtained from Institute of Plant Protection, Chinese Academy of Agriculture Sciences. The susceptible strain (S) was maintained on an artificial diet [25] under Bt toxin-free conditions to preserve baseline susceptibility. Resistance levels to Bt toxins were determined through 7-day diet-incorporated dose–response bioassays for AbR,

AcR, FR, and IeR strains, respectively. The AbR, AcR, FR, and IeR strains have respectively tested more than 710-, 400-, 500-, and 800-fold resistance ratios under bioassays. All colonies were reared in a controlled climate chamber maintained at 27 ± 1 °C, 70–80% relative humidity, under a 16:8 h (light/dark) photoperiod.

2.2. Dissection of Gut Tissues and Extraction of DNA

The fifth instar ACB larvae were surfaced-sterilized by immersion in 70% ethanol for 3 min, followed by three saline washes (5 s each) to remove residual ethanol. The larvae were first immobilized via cold anesthesia (5 min on crushed ice) before dissection in a laminar flow hood. Midgut tissues were harvested using sterilized microsurgical tools in sterile saline solution and immediately flash-frozen in liquid nitrogen and stored at −80 °C until genomic DNA extraction. Each pooled sample was collected from the midguts of fifty-fifth-instar larvae representing different strains. Genomic DNA was extracted from dissected midgut tissues using a TGuide S96 Magnetic Soil/Stool DNA Kit (Tiangen Biotech Beijing Co., Ltd., Beijing, China) in accordance with the manufacturer's protocol. DNA concentrated samples were quantified using the Qubit dsDNA HS Assay Kit and Qubit 4.0 Fluorometer (Thermo Fisher Scientific, Eugene, OR, USA) [26], and DNA integrity and fragment size distribution were evaluated by electrophoresis on a 1% (*w*/*v*) agarose gel stained with GeneGreen Nucleic Acid Dye (Tiangen Biotech Beijing Co., Ltd., Beijing, China), with visualization under UV light.

2.3. Amplification of 16S rRNA Gene Sequences for Microbial Community

The V3–V4 region of 16S rRNA gene from the ACB midgut microbial community was amplified using universal primers (338F: 5′-ACTCCTACGGGAGGCAGCA; and 806R: 5′-GGACTACNNGGGTATCTAAT). PCR amplifications were set up in a 10 μL reaction system that contained a 10 ng template, 0.2 μM of each primer, 1× loading buffer, and nuclease-free water. Target bands were purified using AgencourtAMPure XP Beads (Beckman Coulter, Indianapolis, IN, USA) and quantified using the Qubit dsDNA HS Assay Kit on a Qubit 4.0 Fluorometer (Invitrogen, Thermo Fisher Scientific, Eugene, OR, USA). Libraries were prepared and sequenced on an Illumina NovaSeq 6000 platform (Illumina, Santiago, CA, USA), generating 250 bp paired-end reads [26].

2.4. PCR-Amplified DNA Sequencing

Following total DNA extraction from all the midgut tissues, specific primers were designed based on conserved region sequences. Illumina-compatible sequencing adapters were incorporated into the primers tails to enable downstream library preparation. The Amplification PCR products underwent purification, quantified via fluorometry (Qubit 4.0), and normalized to equimolar concentrations (10 nM) for pooled library construction. The qualified libraries were sequenced with Illumina HiSeq 2500. Raw image data files were converted into sequence reads through base calling. The optimized sequences were filtered and pair-ended merged to obtain optimized sequences (Tags). The optimized sequences were clustered and assigned to operational taxonomic units (OTUs), and species classification was performed according to the sequence composition of each OTU [27]. The number of sequences was statistically processed to assess the quality of the data. The tags were clustered at 97% similarity level using UCLUST in QIIME (version 2.0.0) software to identify OTUs [28]. OTUs were annotated based on the Silva (bacteria) and UNITE (fungal) taxonomic databases.

2.5. Midgut Bacteria Diversity Analysis

The taxonomic information of each OTU was obtained by comparing the representative sequences of OTUs with the microbial reference database. This enabled the quantitative analysis of microbial community composition across taxonomic levels (phylum, class, order, family, genus, and species) for each sample. Using QIIME software, species abundance tables were generated at different levels, and community structure maps were created at different taxonomic levels using the R language tool.

2.6. Isolation and Characterization of *Enterococcus* and *Klebsiella* Species

The genera *Enterococcus* and *Klebsiella* were found to be highly dominant in Bt-resistant strains. To isolate these bacterial genera, 5th-instar larvae of ACB were first immersed in 70% ethanol for 3 min to eliminate surface bacteria. Subsequently, the midgut was dissected under sterile conditions and collected in sterile centrifuge tubes containing 20 µL of sterile water. The gut contents were then homogenized using sterile grinding pestles. The resulting liquid was streaked onto MIAC medium plates (Qingdao Hi-Tech Industrial Park Hope Bio-Technology Co., Ltd., Qingdao, China), which are specifically designed for the cultivation of *Klebsiella* species and cultured at 30 °C for 24 h. For the isolation of *Enterococcus* species, the gut liquid was streaked onto bile aesculin azide agar plates (Qingdao Hi-Tech Industrial Park Hope Bio-Technology Co., Ltd., Qingdao, China). The experiment was conducted with three biological replicates. Bacteria exhibiting identical morphologies were selected for subculture on the corresponding medium plates.

For the identification of *Enterococcus* species, 10 bacterial colonies were chosen for PCR amplification using the 16S rRNA gene with universal primers Ent-27F (5′-AGAGTTTGATCCTGGCTCAG-3′) and Ent-1492R (5′-TACGGTTACCTTGTTACGACTT-3′). Similarly, for the identification of *Klebsiella* species, 10 bacterial colonies were selected and amplified with specific primers Kle-F: TGGCCCGCGCCCAGGGTTCGAAA and Kel-R: GATGTCGTCATCGTTGATGCCGAG. PCR products were characterized by Sangon Biotech Co., Ltd. (Shanghai, China) and then subjected to BLAST (version 2.14.0) searches against the National Center for Biotechnology Information (NCBI) database for identification. Phylogenetic analysis was performed using the neighbor-joining method in MEGA 11 software.

2.7. Virulence Assay of *E. faecalis* and Cry1Ab Susceptibility After Levofloxacin Treatment

The virulence of *E. faecalis* on ACB larvae was assessed using feeding and injection methods. For the feeding assay, 50 mL of *E. faecalis* solution in logarithmic growth stage was centrifuged at 4000 rpm for 10 min to discard supernatant. Then, we blended it with LB liquid medium to OD600 = 1.0 solution and incorporated it into an artificial diet with five bacteria concentration gradients, which with a normal diet served as the control. Each treatment included 24 larvae and was repeated three times. For the injection assay, enterococcal solution at the logarithmic growth stage was centrifuged at 3500 rpm for 15 min, and the supernatant was discarded. Then, we blended it with deionized water to OD600 = 1.0 solution, which was injected 1 µL into 3 instar ACB larval hemolymph while injecting deionized water as the control group.

The role of *E. faecalis* in Cry1Ab resistance was investigated by supplementing artificial diet with levofloxacin (1 mg/mL). After feeding ACB larvae with the levofloxacin-supplemented diet for 24 h, the larvae were divided into two groups: one group was transferred to an artificial diet containing Cry1Ab protein (LC50 = 6.28 ng/cm^2), and the other was transferred to a diet containing both Cry1Ab protein and *E. faecalis*. Treatment groups (n = 24 larvae each) were independently replicated in triplicate.

GraphPad Prism 9.5.0 was used to statistically analyze the results of virulence of *E. faecalis* and Cry1Ab protein virulence bioassay on ACB larvae, while the t-test and one-way ANOVA were used to determine the statistical significance.

3. Results

3.1. Overview of the 16s-RNA Sequencing Data

The V3-V4 region of 16S rRNA was amplified and sequenced from five samples, generating 438,890–746,708 raw reads. After quality control and chimeric removal, 391,221 to 664,608 high-quality tags were retained (Table 1). The sequences clustered into 1640–2754 OTUs per sample at 97% similarity.

Table 1. Statistics of different ACB strains sequencing data.

Strain IDs	PE Reads	Raw Tags	Clean Tags	Effective Tags	AvgLen (bp)	GC (%)	Q20 (%)	Q30 (%)	Effective (%)
AbR	476,918	446,887	400,958	399,519	425	53.58	94.84	90.54	83.77
AcR	720,884	676,881	603,479	600,280	428	52.89	94.73	90.37	83.27
FR	793,527	746,708	666,106	664,608	429	52.77	94.79	90.51	83.75
IeR	696,613	654,749	586,431	583,344	427	53.10	94.82	90.52	83.74
S	467,691	438,890	393,593	391,221	426	53.23	94.89	90.62	83.65

3.2. Different Gut Bacterial Communities in BtS and BtR ACB Strains

3.2.1. The Relative Abundance of Species at the Genus Level

Dominant bacterial genera across the five strains were *Enterococcus* (41–83%), *Klebsiella* (0.1–19%), and *Bacteroides* (3–12%) (Figure 1), while *Enterococcus* exhibited the highest relative abundance in Bt-resistant strains.

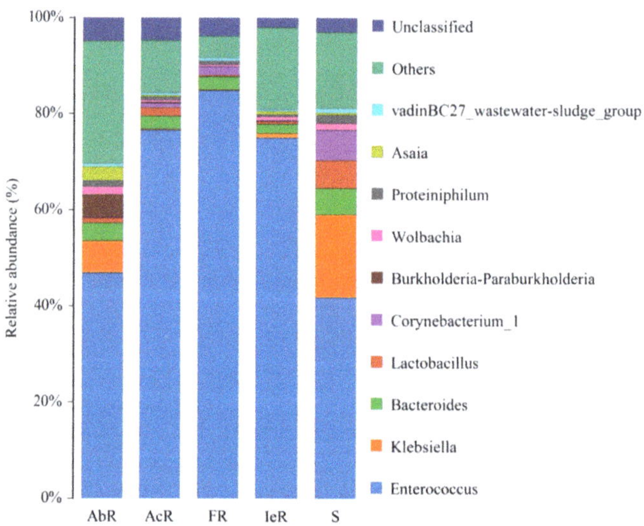

Figure 1. Relative abundance of the top 12 of midgut bacteria genera from BtR (AbR, AcR, FR, and IeR) and BtS (S) strains of Asian corn borer.

The *Enterococcus* abundance in ACB resistant strains (45% for AbR, 73% for AcR, 83% for FR, and 70% for IeR), as well as the susceptible strain (41% for S) were compared. The relative abundance of *Lactobacillus*, *Klebsiella*, and *Bacteroides* were significantly higher in Bt-susceptible strains than in Bt-resistant strains.

3.2.2. Isolation and Identification of Microorganisms

PCR amplification was performed on 10 randomly selected bacterial colonies, successfully identifying *Klebsiella* and *Enterococcus* species. The target fragment size for *Klebsiella* was approximately 300 bp, while for *Enterococcus* was approximately 1300 bp. We added

Enterococcus spp. as positive control. The agarose gel electrophoresis results indicate that the selected bacterial isolate belongs to the *Enterococcus* genus (Figure 2). Phylogenetic analysis using MEGA 11 software identified the isolated the *Enterococcus* strain as *Enterococcus faecalis*.

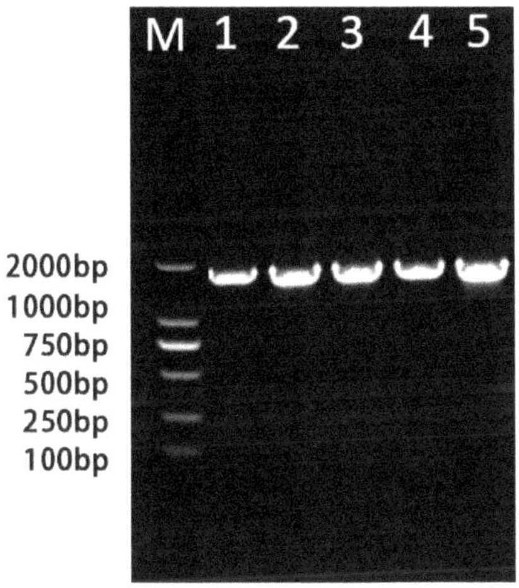

Figure 2. Agarose gel electrophoresis of *Enterococcus*. M: 2000 bp marker. 1–4: Bacterial solution electrophoresis results. 5: Positive control.

3.3. The Impact of Gut Microbiota on Cry1Ab Resistance in Asian Corn Borer

3.3.1. Virulence Evaluation of *Enterococcus faecalis* on ACB Larvae

The results indicated that ACB larvae fed on diets with different *Enterococcus* concentration gradients maintained survival rates above 80%, with no significant differences compared to the control diet. However, larvae injected with the *Enterococcus* solution exhibited significantly reduced survival rates (Figure 3).

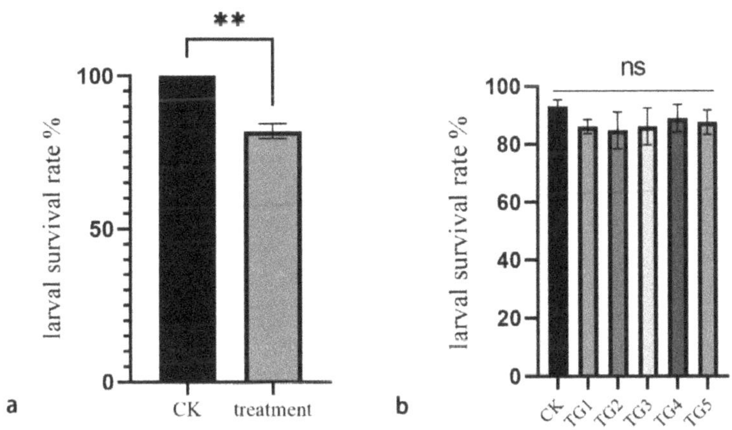

Figure 3. Larval survival rate of Asian corn borer after injection or feeding with *E. faecalis*. (**a**) Asian corn borer larval survival rate after injecting double-distilled water (CK) and *E. faecalis* (treatment). (**b**) Asian corn borer larval survival rate after feeding on normal diet (CK) and *E. faecalis* diet 0.0625 mg/mL (TG1), 0.125 mg/mL (TG2), 0.25 mg/mL (TG3), 0.5 mg/mL (TG4), and 1 mg/mL (TG5). "**": indicates the difference is significant between CK and treatment at 0.01 level; ns: indicates no significant between CK and different treatments.

3.3.2. *E. faecalis* Influence in ACB of Cry1Ab Resistance

The experiment yielded several noteworthy findings. The survival rate of newly hatched larvae continuously fed with levofloxacin was 91.67%, compared to 93.75% in the control group fed with a regular diet, showing no significant difference between the two groups. However, the survival rate of larvae treated with levofloxacin for 24 h and then fed on an artificial diet containing Cry1Ab protein (at a concentration of LC50 = 6.28 ng/cm^2) was significantly higher than that of larvae directly fed on the Cry1Ab-containing artificial diet. When larvae treated with levofloxacin were fed on a diet containing both Cry1Ab protein and *E. faecalis*, their survival rate was 38.54%, which was significantly different from the group Z/C serving as the control group ($p < 0.001$) (Figure 4).

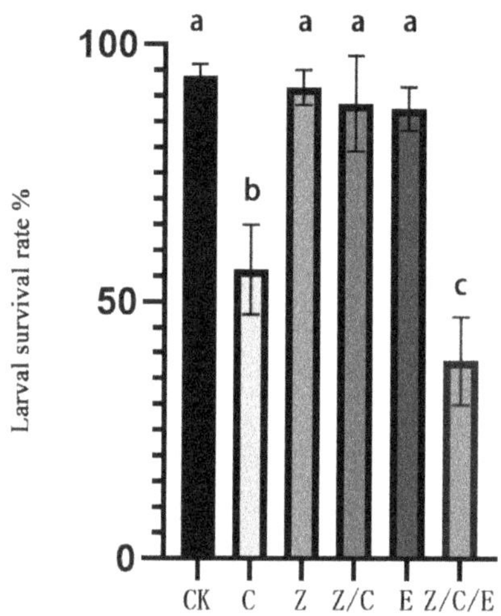

Figure 4. Survival rate of ACB larvae feeding on different treatments with artificial diet. CK: normal artificial diet; C: normal artificial diet containing Cry1Ab protein; Z: normal artificial diet with levofloxacin-supplemented; E. normal artificial diet containing *E. faecalis*; Z/C: normal artificial diet containing Cry1Ab protein and levofloxacin; Z/C/E: normal artificial diet containing Cry1Ab protein, levofloxacin, and *E. faecalis*. Different low case letters above columns indicate statistical differences at $p < 0.01$.

4. Discussion

Different *Bacillus thuringiensis* products have been developed for pest control in agriculture and also against mosquito species [29]. Transgenic Bt plants have been proven effective for controlling Lepidopteran pests. However, since the first Bt resistance news of *Plodia interpunctella* occurred in 1985 [30], many more similar cases were reported. These findings raise concerns about the long-term efficacy of Bt toxins and pose a significant challenge to maintaining the effectiveness of both Bt-based pesticides and gene-modified plants expressing Bt toxins. *Plutella xylostella* was the only insect to eventually develop resistance to Bt applied as a biopesticide in previous research [31]. However, laboratory selection experiments revealed that over 50% of tested moth species (Noctuidae, Plutellidae, and Pyralidae) developed >10-fold resistance to Bt toxins, underscoring the widespread evolutionary capacity of Lepidoptera to adapt to Bt pressure [32]. This highlights an urgent need to decipher resistance mechanisms beyond traditional explanations (e.g., receptor mutations), including the overlooked role of gut microbiota.

Recent evidence suggests that the insect midgut significantly modulates host susceptibility to Bt toxins. Dominant bacterial genera in Lepidopteran midguts, such as *Pseudomonas*, *Bacillus*, *Enterobacter*, and *Enterococcus*, are not only vital for nutrient assimilation but also promote detoxification of plant secondary metabolites, potentially altering toxin efficacy [33,34]. After feeding gypsy moth larvae with different concentrations of antibiotics to reduce the midgut bacteria, the mortality of larvae fed Bt toxin was inversely proportional to the antibiotic concentration. This reduction in mortality was accompanied by reduced populations of culturable *Enterococcus* and *Enterobacter* from the midguts of larvae [22]. Bt protein treatment can alter gut microbial community composition in *Spodoptera exigua* compared with the normal populations [35]. Bt toxin exposure caused severe midgut epithelial disruption, which enables gut bacteria to translocate into the hemocoel [36]. This bacterial invasion may convert commensal bacteria into pathogens that potentially accelerate larvae mortality. In a previous study about *Manduca sexta*, Mason et al. [37] found that a common gut microbiota *E. faecalis* invades the hemolymph of *M. sexta* larvae within 48 h post-ingestion. The bacterial load increased progressively until death, though the precise translocation mechanisms remain unknown. This hypothesis was supported by the results presented in our study. Chen et al. [38] demonstrated that indigenous *Enterococcus* spp. synergistically enhanced Cry1Ca-mediated mortality in *Chilo suppressalis* larvae by inducing melanization and hemocyte apoptosis. Therefore, exploring the functional properties of *Enterococcus* in pest midguts could be a critical avenue for future research on of Cry-toxin resistance mechanisms.

In previous research, it was demonstrated that Bt proteins induce the collapse of the membrane, subsequently facilitating the invasion of midgut microbiota into the hemolymph and ultimately resulting in insect death due to sepsis [3]. In this study, ACB larvae fed on *E. faecalis* diet showed no significant difference in survival compared to those on a normal diet. However, direct midgut injection of *E. faecalis* substantially altered larval survival (Figure 3). Notably, when gut bacteria (including *E. faecalis*) were depleted using a levofloxacin-containing diet, ACB survival rates on Cry1Ab-treated feed significantly increased (Figure 4). These results demonstrate that *E. faecalis* enhances ACB susceptibility to Cry1Ab and suggest that gut microbiota composition critically modulates larval resistance to Bt toxins. This outcome aligns with the results of prior studies, which reported that a reduction in the gut microbiota of the diamondback moth leads to a decreased sensitivity of its larvae to the Cry1Ac protein [39]. Although the mechanism by which ACB reduces *E. faecalis* remains unknown, hemocytes from *Heliothis virescens* larvae exposed to Cry1Ac toxin exhibited upregulated expression of immune genes associated with Bt intoxication, including antimicrobial peptides, cytokines, and protease inhibitors [40]. We hypothesize that Cry1Ab-resistant ACB strains mount a similar immune response to suppress *E. faecalis* in the midgut.

In conclusion, this study reveals that the larval midgut bacteria participate in the death mechanism of Cry toxins in *O. furnacalis*. The microbial community composition and richness in Bt-resistant and Bt-susceptible strains were significantly different. We hypothesize that the midgut bacteria may influence Bt toxin resistance in many pests, though this requires experimental validation in future studies. The implications are significant for Bt-gene modified corn cultivation and may lead to new strategies for biological pest control.

Author Contributions: Conceptualization, Y.Z., T.Z. and J.Z.; methodology, Y.Z. and T.Z.; validation, Y.Z., T.Z. and J.Z.; formal analysis, X.L., Z.Z. and J.Z.; investigation, X.L., Z.Z. and J.Z.; resources, Y.Z. and T.Z.; data curation, Z.Z. and J.Z.; writing—original draft preparation, J.Z.; writing—review and editing, T.Z. and J.Z.; supervision, Y.Z. and T.Z.; project administration, T.Z.; funding acquisition, T.Z. All authors have read and agreed to the published version of the manuscript.

Funding: This work was financially supported by the Innovation Program of Chinese Academy of Agricultural Sciences (CAAS-CSCB-202303).

Data Availability Statement: The original contributions presented in this study are included in the article. Further inquiries can be directed to the corresponding author.

Conflicts of Interest: The authors declare no conflicts of interest.

References

1. Schnepf, E.; Crickmore, N.; Van Rie, J.; Lereclus, D.; Baum, J.; Feitelson, J.; Zeigler, D.R.; Dean, D.H. *Bacillus thuringiensis* and Its Pesticidal Crystal Proteins. *Microbiol. Mol. Biol. Rev.* **1998**, *62*, 775–806. [CrossRef] [PubMed]
2. Bravo, A.; Gill, S.S.; Soberon, M. Mode of Action of *Bacillus thuringiensis* Cry and Cyt Toxins and Their Potential for Insect Control. *Toxicon* **2007**, *49*, 423–435. [CrossRef]
3. Gill, S.S.; Cowles, E.A.; Pietrantonio, P.V. The Mode of Action of *Bacillus thuringiensis* Endotoxins. *Annu. Rev. Entomol.* **1992**, *37*, 615–636. [CrossRef]
4. Tabashnik, B.E.; Carriere, Y. Surge in Insect Resistance to Transgenic Crops and Prospects for Sustainability. *Nat. Biotechnol.* **2017**, *35*, 926–935. [CrossRef]
5. Li, J.; Zhang, L. Current Understandings of Olfactory Molecular Events in the Asian Corn Borer, *Ostrinia furnacalis* (Lepidoptera: Crambidae). *Arch. Insect Biochem. Physiol.* **2023**, *112*, e21996. [CrossRef]
6. Nafus, D.M.; Schreiner, I.H. Review of the Biology and Control of the Asian Corn Borer, *Ostrinia furnacalis* (Lep: Pyralidae). *Trop. Pest Manag.* **1991**, *37*, 41–56. [CrossRef]
7. He, K.L.; Wang, Z.Y.; Zhou, D.R.; Wen, L.P.; Song, Y.Y.; Yao, Z.Y. Evaluation of Transgenic Bt Corn for Resistance to the Asian Corn Borer (Lepidoptera: Pyralidae). *J. Econ. Entomol.* **2003**, *96*, 935–940. [CrossRef]
8. Wang, Y.; Wang, Y.; Wang, Z.; Bravo, A.; Soberon, M.; He, K. Genetic Basis of Cry1F-Resistance in a Laboratory Selected Asian Corn Borer Strain and Its Cross-Resistance to Other *Bacillus thuringiensis* Toxins. *PLoS ONE* **2016**, *11*, e0161189. [CrossRef]
9. Wang, Y.; Yang, J.; Quan, Y.; Wang, Z.; Cai, W.; He, K. Characterization of Asian Corn Borer Resistance to Bt Toxin Cry1Ie. *Toxins* **2017**, *9*, 186. [CrossRef]
10. Wang, D.; Wang, Z.; He, K.; Cong, B.; Bai, S.; Wen, L. Temporal and Spatial Expression of Cry1Ab Toxin in Transgenic Bt Corn and Its Effects on Asian Corn Borer, *Ostrinia furnacalis* (Guenee). *Sci. Agric. Sin.* **2004**, *37*, 1155–1159.
11. Jin, T.; Chang, X.; Gatehouse, A.M.R.; Wang, Z.; Edwards, M.G.; He, K. Downregulation and Mutation of a Cadherin Gene Associated with Cry1Ac Resistance in the Asian Corn Borer, *Ostrinia furnacalis* (Guenee). *Toxins* **2014**, *6*, 2676–2693. [CrossRef] [PubMed]
12. Xiao, Y.; Wu, K. Recent Progress on the Interaction between Insects and *Bacillus thuringiensis* Crops. *Philos. Trans. R. Soc. B-Biol. Sci.* **2019**, *374*, 20180316. [CrossRef]
13. Pinos, D.; Wang, Y.; Hernandez-Martinez, P.; He, K.; Ferre, J. Alteration of a Cry1A Shared Binding Site in a Cry1Ab-Selected Colony of *Ostrinia furnacalis*. *Toxins* **2022**, *14*, 32. [CrossRef]
14. Tanaka, S.; Miyamoto, K.; Noda, H.; Jurat-Fuentes, J.L.; Yoshizawa, Y.; Endo, H.; Sato, R. The ATP-Binding Cassette Transporter Subfamily C Member 2 in Bombyxmori Larvae Is a Functional Receptor for Cry Toxins from *Bacillus thuringiensis*. *FEBS J.* **2013**, *280*, 1782–1794. [CrossRef]
15. Knight, P.J.; Crickmore, N.; Ellar, D.J. The Receptor for *Bacillus thuringiensis* CrylA(c) Delta-Endotoxin in the Brush Border Membrane of the Lepidopteran Manduca Sexta Is Aminopeptidase N. *Mol. Microbiol.* **1994**, *11*, 429–436. [CrossRef]
16. Luo, K.; Sangadala, S.; Masson, L.; Mazza, A.; Brousseau, R.; Adang, M.J. The Heliothis Virescens 170kDa Aminopeptidase Functions as "Receptor a" by Mediating Specific *Bacillus thuringiensis* Cry1A δ-Endotoxin Binding and Pore Formation. Insect Biochem. *Mol. Biol.* **1997**, *27*, 735–743. [CrossRef]
17. Vadlamudi, R.K.; Weber, E.; Ji, I.H.; Ji, T.H.; Bulla, L.A. Cloning and Expression of a Receptor for an Insecticidal Toxin of *Bacillus-thuringiensis*. *J. Biol. Chem.* **1995**, *270*, 5490–5494. [CrossRef]
18. Morin, S.; Biggs, R.W.; Sisterson, M.S.; Shriver, L.; Ellers-Kirk, C.; Higginson, D.; Holley, D.; Gahan, L.J.; Heckel, D.G.; Carrière, Y.; et al. Three Cadherin Alleles Associated with Resistance to *Bacillus thuringiensis* in Pink Bollworm. *Proc. Natl. Acad. Sci. USA* **2003**, *100*, 5004–5009. [CrossRef]
19. Zhang, X.; Zhang, F.; Lu, X. Diversity and Functional Roles of the Gut Microbiota in Lepidopteran Insects. *Microorganisms* **2022**, *10*, 1234. [CrossRef]
20. Zhu, Y.; Zhang, Y.; Wang, X.; Yin, Y.; Du, Y. *Wolbachia* Modify Host Cell Metabolite Profiles in Response to Short-term Temperature Stress. *Environ. Microbiol. Rep.* **2024**, *16*, e70013. [CrossRef]
21. Engel, P.; Moran, N.A. The Gut Microbiota of Insects—Diversity in Structure and Function. *FEMS Microbiol. Rev.* **2013**, *37*, 699–735. [CrossRef] [PubMed]
22. Broderick, N.A.; Raffa, K.F.; Handelsman, J. Midgut Bacteria Required for *Bacillus thuringiensis* Insecticidal Activity. *Proc. Natl. Acad. Sci. USA* **2006**, *103*, 15196–15199. [CrossRef]

23. Paddock, K.J.; Pereira, A.E.; Finke, D.L.; Ericsson, A.C.; Hibbard, B.E.; Shelby, K.S. Host Resistance to *Bacillus thuringiensis* Is Linked to Altered Bacterial Community within a Specialist Insect Herbivore. *Mol. Ecol.* **2021**, *30*, 5438–5453. [CrossRef]
24. Orozco-Flores, A.A.; Valadez-Lira, J.A.; Oppert, B.; Gomez-Flores, R.; Tamez-Guerra, R.; Rodriguez-Padilla, C.; Tamez-Guerra, P. Regulation by Gut Bacteria of Immune Response, *Bacillus thuringiensis* Susceptibility and Hemolin Expression in Plodia Interpunctella. *J. Insect Physiol.* **2017**, *98*, 275–283. [CrossRef]
25. Rahayu, T.; Trisyono, Y.A. Witjaksono Fitness of Asian Corn Borer, *Ostrinia Furnacalis* (Lepidoptera: Crambidae) Reared in an Artificial Diet. *J. Asia-Pac. Entomol.* **2018**, *21*, 823–828. [CrossRef]
26. Xu, T.; Wang, Y.; Wang, Y.; Bi, S.; Hu, B.; Hu, F.; Xu, L. Comparison of Gut Microbial Community between Bt-Resistant and Susceptible Strains of *Ostrinia Furnacalis*. *Agronomy* **2023**, *13*, 1923. [CrossRef]
27. Govindharaj, G.-P.-P.; Choudhary, J.S.; Panda, R.M.; Basana-Gowda, G.; Annamalai, M.; Patil, N.; Khan, R.M.; Banra, S.; Srivastava, K.; Mohapatra, S.D. Bacterial Communities in *Nilaparvata lugens* (Stål) (Hemiptera: Delphacidae) Showed Significant Variation among the Developmental Stages with Functional Diversity. *Heliyon* **2025**, *11*, e42776. [CrossRef]
28. Bolyen, E.; Rideout, J.R.; Dillon, M.R.; Bokulich, N.A.; Abnet, C.C.; Al-Ghalith, G.A.; Alexander, H.; Alm, E.J.; Arumugam, M.; Asnicar, F.; et al. Reproducible, Interactive, Scalable and Extensible Microbiome Data Science Using QIIME 2. *Nat. Biotechnol.* **2019**, *37*, 852–857. [CrossRef]
29. Bravo, A.; Likitvivatanavong, S.; Gill, S.S.; Soberon, M. *Bacillus thuringiensis*: A Story of a Successful Bioinsecticide. *Insect Biochem. Mol. Biol.* **2011**, *41*, 423–431. [CrossRef]
30. McGaughey, W.H. Insect Resistance to the Biological Insecticide *Bacillus thuringiensis*. *Science* **1985**, *229*, 193–195. [CrossRef]
31. Ferré, J.; Van Rie, J. Biochemistry and Genetics of Insect Resistance to *Bacillus thuringiensis*. *Annu. Rev. Entomol.* **2002**, *47*, 501–533. [CrossRef] [PubMed]
32. Tabashnik, B.E. Evolution of Resistance to *Bacillus thuringiensis*. *Annu. Rev. Entomol.* **1994**, *39*, 47–79. [CrossRef]
33. Sabree, Z.L.; Kambhampati, S.; Moran, N.A. Nitrogen Recycling and Nutritional Provisioning by *Blattabacterium*, the Cockroach Endosymbiont. *Proc. Natl. Acad. Sci. USA* **2009**, *106*, 19521–19526. [CrossRef]
34. Voirol, L.R.P.; Frago, E.; Kaltenpoth, M.; Hilker, M.; Fatouros, N.E. Bacterial Symbionts in Lepidoptera: Their Diversity, Transmission, and Impact on the Host. *Front. Microbiol.* **2018**, *9*, 556. [CrossRef]
35. Li, Y.; Zhao, D.; Wu, H.; Ji, Y.; Liu, Z.; Guo, X.; Guo, W.; Bi, Y. Bt GS57 Interaction with Gut Microbiota Accelerates *Spodoptera exigua* Mortality. *Front. Microbiol.* **2022**, *13*, 835227. [CrossRef]
36. Deguenon, J.M.; Dhammi, A.; Ponnusamy, L.; Travanty, N.V.; Cave, G.; Lawrie, R.; Mott, D.; Reisig, D.; Kurtz, R.; Roe, R.M. Bacterial Microbiota of Field-Collected *Helicoverpa zea* (Lepidoptera: Noctuidae) from Transgenic Bt and Non-Bt Cotton. *Microorganisms* **2021**, *9*, 878. [CrossRef]
37. Mason, K.L.; Stepien, T.A.; Blum, J.E.; Holt, J.F.; Labbe, N.H.; Rush, J.S.; Raffa, K.F.; Handelsman, J. From Commensal to Pathogen: Translocation of Enterococcus Faecalis from the Midgut to the Hemocoel of *Manduca sexta*. *mBio* **2011**, *2*, e00065.11. [CrossRef]
38. Chen, G.; Li, Q.; Zhang, C.; Zhao, W.; Jurat-Fuentes, J.L.; Zhou, X.; Chen, F.; Yang, X.; Han, L. Synergism of Cry1Ca Toxicity by Gut Resident *Enterococcus* spp. in the Rice Stem Borer, *Chilo suppressalis*. *Int. J. Biol. Macromol.* **2024**, *257*, 128654. [CrossRef]
39. Li, S.; Xu, X.; De Mandal, S.; Shakeel, M.; Hua, Y.; Shoukat, R.F.; Fu, D.; Jin, F. Gut Microbiota Mediate *Plutella xylostella* Susceptibility to Bt Cry1Ac Protoxin Is Associated with Host Immune Response. *Environ. Pollut.* **2021**, *271*, 116271. [CrossRef]
40. Perera, O.P.; Shelby, K.S.; Popham, H.J.R.; Gould, F.; Adang, M.J.; Jurat-Fuentes, J.L. Generation of a Transcriptome in a Model Lepidopteran Pest, *Heliothis virescens*, Using Multiple Sequencing Strategies for Profiling Midgut Gene Expression. *PLoS ONE* **2015**, *10*, e0128563. [CrossRef]

insects

MDPI

Article

Occurrence and Genetic Variation of *Monolepta hieroglyphica* (Motschulsky, 1858) (Coleoptera: Chrysomelidae), a Newly Emerging Pest, Among Hosts in Northeast China

Wei Sun, Xiuhua Zhang, Jiachun Zhou and Yuebo Gao *

Northeast Agricultural Research Center of China, Institute of Plant Protection, Jilin Academy of Agricultural Sciences, Gongzhuling 136100, China; swswsw1221@sina.com (W.S.); hshs922922@163.com (X.Z.); pxzkn1130@163.com (J.Z.)
* Correspondence: gaoyuebo8328@163.com

Simple Summary: Northeastern China is recognized as a crucial grain-producing region, but food security is severely affected by diverse pests. Due to changes in climate, cultivation patterns, and crop distribution, the leaf beetle *Monolepta hieroglyphica* (Motschulsky, 1858) (Coleoptera: Chrysomelidae) has emerged as a destructive pest. However, its occurrence across different hosts remains poorly understood. This study analyzed the pest's occurrence patterns and genetic diversity through systematic observation and mitochondrial DNA markers. These findings are essential for developing effective pest control strategies in the region.

Abstract: The northeast region of China plays a crucial role in crop production. The leaf beetle *Monolepta hieroglyphica* (Motschulsky, 1858) (Coleoptera: Chrysomelidae) has emerged as a potential threat to food security in the region. With a wide distribution spanning Asia and Russia, this beetle affects various crops. However, limited information is available regarding its occurrence patterns and genetic diversity among major crops in the region. Based on systematic observations across various hosts, coupled with genetic variation analysis using mitochondrial DNA markers, the main results were as follows. Leaf beetle occurrence varied among hosts, peaking from late July to mid-August, with maize and soybean fields exhibiting higher infestation rates compared with other crops. Notably, late-cultivated maize fields harbored the highest beetle numbers due to the species' preference for young leaves. The host transfer trajectory may have originated in soybean and weeds, with subsequent alternation between host plants and other crops, before the final migration to cabbage and late-cultivated maize fields. Genetic analysis revealed nine COI haplotypes, four COII haplotypes, eleven Cytb haplotypes, and twenty-one combined haplotypes. No clear relationship existed between genetic diversity and occurrence, and no distinct host-based genetic patterns emerged from neighbor-joining tree and haplotype network analyses. High gene flow rates were observed, likely contributing to decreased genetic variation. An analysis of molecular variance results indicated major genetic variation within populations, although genetic distance and haplotype distribution indicated divergence among host populations. These results provide foundational data for developing effective *M. hieroglyphica* pest management strategies.

Keywords: leaf beetle; host population; spatial dynamics; dispersal; molecular marker; genetic diversity

1. Introduction

Northeastern China, including Eastern Inner Mongolia, Jilin, Liaoning, and Heilongjiang Provinces, is regarded as the country's largest grain production base. With large plain topography, this region benefits from an April–October growing season, and harsh winters limit insect activity. Major crops include maize, soybean, and rice, with other crops such as sunflower, wheat, millet, peanut, and sorghum cultivated in certain areas. The region experiences substantial agricultural losses due to various pests, including the oriental armyworm *Mythimna separata* (Walker) (Lepidoptera: Noctuidae) and aphids [1,2]. However, due to changes in climate, cultivation patterns, and crop allocation, the leaf beetle *Monolepta hieroglyphica* (Motschulsky, 1858) (Coleoptera: Chrysomelidae) has emerged as a new threat to food security, particularly affecting maize and soybean crops [3,4].

This pest is widely distributed across East Asia, Southeast Asia, and Russia [5]. In China, this species exhibits a broad provincial distribution, with overwintering occurring in the egg stage [3]. This leaf beetle is a polyphagous pest, feeding on a wide variety of crop and weed species, with larvae and adults directly inflicting crop damage [6]. The larvae are an underground pest of crop plants, whereas the adults damage leaves, flowers, filaments, pollen, floral organs, clusters, and grains, and may negatively affect pollination [7]. Adults possess wings, allowing some mobility, including short-distance dispersal (2–5 m) [6].

The economic impact of the pest has prompted extensive research into its occurrence patterns, phylogenetics, insecticidal mechanisms, drip irrigation control, and biological control [4,8–10]. Studies on *M. hieroglyphica* genetics have primarily examined its complete mitochondrial genome, molecular systematics, and molecular markers [3,11–14]. However, little attention has been paid to the pest's occurrence among different hosts, resulting in limited understanding of host transfer migration. Furthermore, clear differentiation exists among *M. hieroglyphica* geographic populations in northern China [13], indicating population-level variation. Similarly, morphology and biological differences exist among host populations, such as variations in adult size related to emergence periods [7] and discrepancies in life history and reproduction [15]. However, it remains unclear whether these differences are due to host specialization. This scientific question warrants further investigation.

This study explores the spatial dynamics and genetic variations of the leaf beetle across different crops in Northeast China. Understanding occurrence patterns and genetic diversity is pivotal for devising effective pest management strategies. A key aspect of this study lies in the experimental site selection, which included all host plant species within a relatively small area, eliminating topographical, temporal, geographic, and climatic differences that could impede the pest's dispersal. Another innovative aspect involved systemic field surveys combined with molecular markers, proving valuable for studying host genetic variation. The study's findings provide unique regional insights into pest management.

2. Materials and Methods

2.1. Population Dynamics Analyses

Field surveys were conducted in Gongzhuling (43°32′09″ N, 124°49′28″ E), situated in the central agricultural plain of Jilin Province, during the periods of June to October in 2022 and 2023. In total, 11 host fields were established, including the major crops in Northeast China. The experimental site and sampling details are provided in Table 1. Weed species primarily belonged to the grass family. Planting in most host fields occurred during April and May. Cabbage seed planting was performed on 12 August 2022 and 2 August 2023, aligning with general practices in the local area. Late-cultivated maize [maize (L)] seeds were planted on 2 July 2022 and 30 June 2023.

Table 1. Information regarding the experimental site and molecular samples.

Population Code	Host	Sampling Methods			Molecular Samples	PCR-Positive Samples				Sampling Date		Harvest Date	
		Y	S	O		COI	COII	Cytb	COM	2022	2023	2022	2023
ZC	weed	-	✓	-	40	36	40	40	36	06-01~10-17	06-01~10-16	-	-
DD	soybean	✓	✓	-	40	38	39	35	35	06-01~10-17	06-01~10-16	09-25	09-22~09-30
BC	cabbage	✓	-	✓	30	29	30	30	29	09-08~10-17	08-22~10-16	10-10	10-01
YM	maize	✓	-	✓	40	38	40	39	38	06-01~10-17	06-01~10-16	10-12	10-07
YW	maize (L)	✓	-	✓	12	11	12	12	11	07-16~10-17	07-13~10-16	10-12	10-07
GZ	millet	✓	-	✓	30	25	30	28	25	06-01~10-17	06-01~10-16	10-04	10-11
GL	sorghum	✓	-	✓	30	29	29	29	29	06-01~10-17	06-01~10-16	10-04	10-06
XR	sunflower	✓	-	✓	30	29	30	30	29	06-01~10-17	06-01~10-16	08-15~09-30	08~18-09~25
HS	peanut	✓	-	✓	55	35	44	38	34	06-01~10-17	06-01~10-16	09-06~10-07	09~04-09~28
XM	wheat	✓	-	✓	0	0	0	0	0	06-01~07-17	06-01~07-16	07-20	07-19
SD	rice	✓	-	✓	30	29	30	29	29	06-01~10-17	06-01~10-16	10-02	10-03

Y: yellow traps; S: sweep sampling; O: visual observations; -: no data collected; ✓: this method was used; maize (L): late-cultivated maize.

Field observations were conducted using sweep sampling, visual observations, and yellow traps (Table 1). Sampling methods were adjusted for specific host fields based on their unique characteristics. Sweep sampling is a common method used to estimate the relative abundance of insect communities. Following O'Neill et al. [16] and Whipple et al. [17], 200 random sweeps with a 40 cm diameter sweep net were performed per field. Sweep sampling was used for low-density plant fields, such as weeds and soybean fields. For other host fields, visual observations were performed during random 200 m walking surveys in each field. Notably, field observations in maize (L) were conducted over 120 m from July to October 2022, and over 180 m from July to October 2023. In contrast, cabbage field observations covered 120 m from September to October 2022, and 150 m from August to October 2023. Yellow traps, which effectively record pests' initial and last appearance, were used as supplementary tools. Two yellow traps (20 × 40 cm) were placed at the center of each field, with data recorded every 3 days June–October in 2022 and 2023.

2.2. Molecular Analyses

Mitochondrial DNA (mtDNA) serves as a valuable molecular marker for assessing population genetic diversity and variation [18–23]. Partial COI, COII, and Cytb fragments of mtDNA were selected for use and amplified using the following primer pairs (Table 2). To ensure consistency across sampling years, 10 host populations of *M. hieroglyphica* were collected in 2022 and 2023 from the aforementioned host fields and stored at −20 °C until processing. Table 1 presents the host population sample sizes used for molecular analysis. Among the 337 samples subjected to PCR amplification, differential gene conservation resulted in varying success rates: 299 samples amplified successfully for COI, 324 for COII, and 310 for Cytb. Owing to low occurrence, samples from wheat were not collected. To ensure accuracy, only samples from yellow traps in maize (L) fields were included in the statistical analysis, accounting for the relatively small sample size. Morphologically, identification was performed by Wei Sun using reference materials [24].

Table 2. Primer information.

Gene	Primer Sequences	Primer Source
COI-F	AAAAATAGATTTTATCTAAGCCTTA	Designed from: NCBI MT178239
COI-R	TATGCTCGAGTATCTACATCTATAC	
COII-F	GAGCATCTCCTTTAATAGAACA	[13]
COII-R	GTATAAATGAGTGATTGGCTCC	
Cytb-F	AATTATGGWTGAYTAATTCGAAC	[13]
Cytb-R	AAATATCATTCAGGTTGAATATG	

All the experimental procedures, including PCR design and sequencing, were conducted by Sangon Biotech (Shanghai) Co., Ltd. In total, 337 samples were used for genomic DNA extraction. Genomic DNA was extracted from a portion of *M. hieroglyphica* adult bodies using a genomic DNA purification kit (Sangon Biotech, Shanghai, China). PCR reaction mixtures contained 1 μL of DNA template, 2.5 μL of Taq buffer (with $MgCl_2$), 1 μL of each primer, 1 μL of dNTP, and 0.2 μL of Taq DNA polymerase enzyme (Sangon Biotech, Shanghai, China) in a 25 μL volume with molecular-grade water. PCR cycling parameters were as follows: initial denaturation at 95 °C for 5 min; 10 cycles of denaturation at 94 °C for 30 s, annealing at 63 °C for 30 s (decreasing by 0.5 °C per cycle), extension at 72 °C for 30 s; 30 cycles of denaturation at 95 °C for 30 s, annealing at 58 °C for 30 s, extension at 72 °C for 30 s; and a final extension at 72 °C for 10 min. All PCR reactions were conducted using an ABI Veriti 96-Well system, and samples with successful PCR amplification were sequenced using the ABI 3730 XL (Applied Biosystems, Foster City, CA, USA).

2.3. Data Analyses

The data from observations/sweep sampling and yellow traps were combined for comprehensive analysis, aiming to provide more accurate and complementary information. Statistical analyses and visualizations were performed using Excel 2010. To provide a comprehensive understanding of genetic variation, COI, COII, and Cytb fragments were analyzed both individually and in combination. Sequence alignment, editing, and haplotype definition were performed using Chromas 1.62, DNAMAN V6, and EditSeq 5.01 software. Haplotypes were deposited in the NCBI Genbank database under accession numbers PP038011–PP038019 and PP056518–PP056532. Nucleotide composition, variable sites, transition/transversion ratios, and haplotype genetic distances were calculated using MEGA 4.0 [25]. A phylogenetic tree [neighbor-joining (NJ)] was constructed with the K-2-P model in MEGA 4.0. DnaSP 5 was used to analyze haplotype number (*H*), haplotype diversity (*Hd*), average number of nucleotide differences (*K*), nucleotide diversity (*Pi*), and gene flow estimates [26]. Haplotype networks were generated using Network 4.6.1.6 with median joining [27]. Analyses of molecular variance (AMOVA) and population genetic distance were performed using Arlequin 3.5.1.2 [28].

3. Results

3.1. Population Dynamics

The population dynamics of *M. hieroglyphica* based on the field-collected data are shown in Figure 1. During 2021 and 2022, in the millet field, leaf beetles first appeared on 10 July and last appeared on 11 September, with peak abundance observed from late July to late August. Similarly, in the sunflower field, the initial appearance occurred on 4 July, with the last sighting on 27 August and peak abundance from late July to mid-August. In the peanut field, leaf beetles first and last appeared on 8 July and 27 August, respectively, with peak abundance from late July to mid-August. In the sorghum field, the first sightings were on 17 July, with the last observation on 11 September and peak abundance from late July to late August. The wheat fields exhibited minimal activity, with only one individual observed. Regarding the maize field, leaf beetles first occurred on 2 July, last occurred on 1 October, and peaked from mid-July to mid-August. In the maize (L) field, the beetles were first observed on 16 July and last sighted on 23 September, with peak abundance occurring from late July to mid-August. The cabbage fields first showed the presence of leaf beetles on 22 August, with the last occurrence on 5 October. Regarding both soybean fields and weeds, the beetles initially appeared on 2 July and were last found on 11 September; however, peak abundance was from mid-July to late August in soybean and from mid-July to early September in weeds.

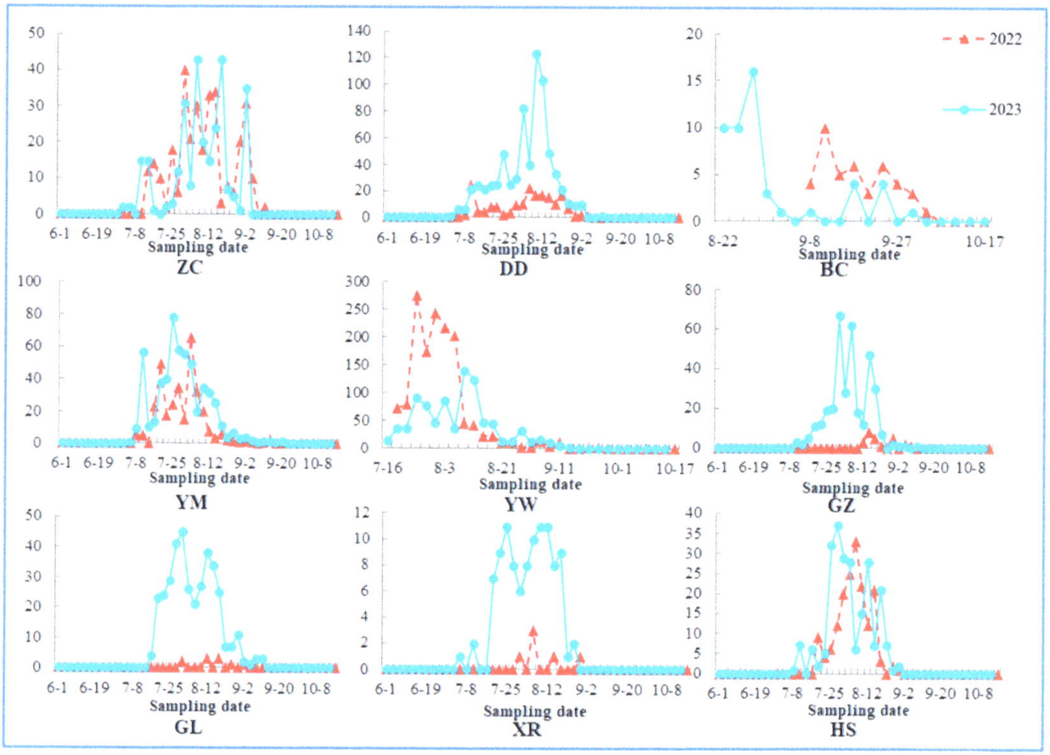

Figure 1. Population dynamics of the leaf beetle among host plant species in 2022 and 2023. The wheat field is excluded owing to low occurrence levels. ZC: weed; DD: soybean; BC: cabbage; YM: maize; YW: maize (L); GZ: millet; GL: sorghum; XR: sunflower; HS: peanut; XM: wheat. The same applies below.

The field observation data from the rice field closely resembled those from other host populations. Given that the study area primarily consists of dry farmland with minimal rice cultivation (not representative of major rice production zones), rice field data were excluded from the statistical analysis to maintain research accuracy. Only molecular study samples were used. No individuals were detected after crop harvesting. After September, no samples were collected in the sunflower and peanut fields, likely due to their earlier harvest. Based on the above observation, the peak occurrence period occurred from late July to mid-August. Combined data from the two sampling years revealed that leaf beetle occurrence commenced earlier in maize, soybean, and weed hosts, whereas in later periods, the pest shifted to maize, maize (L), and cabbage fields.

Despite differences in sampling methods, comparable numbers were obtained (Figure 2). Occurrence rates were relatively higher in 2023. Substantial numbers were observed in the maize (L) field, which made a major contribution to abundance in 2022. Distribution among hosts was more balanced in 2023. In a comprehensive analysis of two-year data, the maize (L) field showed the highest occurrence (2298 individuals) despite the reduced sampling period and distance. The maize (835) and soybean (870) fields exhibited relatively high numbers compared with other crops. The weed (600), sorghum (357), millet (361), and peanut (399) fields also contained numerous beetles, whereas the cabbage fields (92) showed lower numbers, partly attributed to the reduced period and distance. The sunflower fields (101) exhibited fewer individuals, and only one adult was captured in the wheat field.

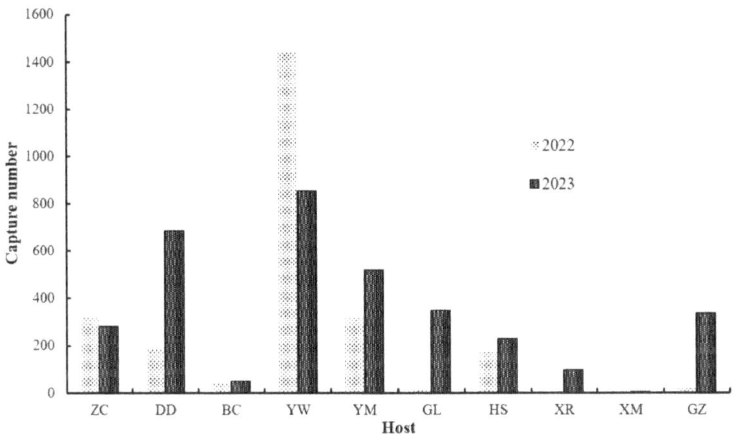

Figure 2. Occurrence levels of the leaf beetle among various hosts.

3.2. Base Composition

The alignment of the COI sequences contained 615 bases with 607 conserved sites and eight single variable sites. The average nucleotide composition was as follows: A: 37.7%; T: 32.8%; C: 15.8%; and G: 13.7%. The transition/transversion ratio (R) was 7. The average A + T content was 70.5%. The alignment of COII sequences, containing 430 bases, had 427 conserved sites and three single variable sites. The average nucleotide composition was as follows: A: 34.9%; T: 41.6%; C: 12.8%; and G: 10.7%. The average A + T content was 76.5%. The alignment of Cytb sequences comprised 430 bases with 423 conserved sites. The sequence included two single variable sites and five parsimony-informative sites. The average nucleotide composition was as follows: A: 40.5%; T: 34.3%; C: 11.3%; and G: 13.9%. The transition/transversion ratio (R) was 12.6. The average A + T content was 74.8%.

The combined COI, COII and Cytb fragment, containing 1475 bases, had 1458 conserved sites and 17 single variable sites. The sequence included nine single variable sites and eight parsimony-informative sites. The average nucleotide composition was as follows: A: 37.7%; T: 35.8%; C: 13.7%; and G: 12.9%. The transition/transversion ratio (R) was 16.4. The average A + T content was 73.5%. The COI, COII, and Cytb fragments were identified with 100% confidence using previously submitted sequences from NCBI (accession nos. MW732714.1). No additions or deletions were observed. Substitutions were predominantly transitions, notably C-T patterns. The high A + T content was consistent with typical insect values.

3.3. Haplotypes

The established network, reflecting haplotype frequencies and distributions (Figure 3), showed no evidence of host plant trends. From 299 individuals, nine unique mtDNA COI haplotypes (C1–C9, NCBI accession nos. PP038011-PP038019, Table S1) were identified. The haplotype content was 3% (9/299). Haplotype C1 was ubiquitous across host populations, constituting 92.64% of individuals. Haplotype C2, the second most frequent haplotype (2.01% of individuals), occurred in four host populations, as did haplotype C4 (1.67% of individuals). Notably, haplotype C4 was consistently present in the maize field across both sampling years. Haplotype C7 was observed in four host populations, representing 1.34% of individuals. The remaining infrequent haplotypes were distributed irregularly among different populations, with the infrequent haplotype C5 detected in soybean and weed populations. The mean genetic distance among the COI haplotypes was 0.003, ranging from 0.002 to 0.003. From 324 individuals, four unique mtDNA COII haplotypes (K1–K4, NCBI accession nos. PP056518-PP056521, Table S1) were identified. The haplotype content was 1.2% (4/324). Haplotype K1, prevalent across all the host populations, accounted for 95.37% of the individuals. Haplotype K3 was found in seven host populations, comprising 3.7% of the individuals. The remaining two haplotypes exhibited irregular distributions among different populations, with haplotype K2 detected in soybean and weed populations and haplotype K4 found in the peanut population. The mean genetic distance among the COII haplotypes was 0.004, ranging from 0.002 to 0.005.

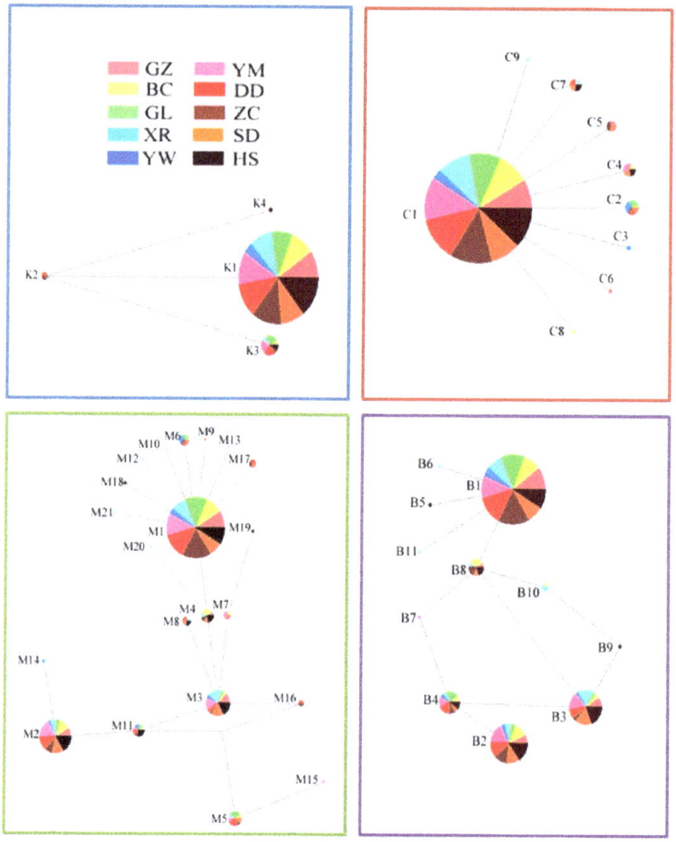

Figure 3. Leaf beetle networks were constructed based on COI, COII, Cytb, and the combined haplotypes. Each haplotype is represented by a circle, with the circle size proportional to the haplotype frequency. Different colors represent different host groups. C1–C9: mtDNA COI haplotypes 1–9. K1–K4: mtDNA COII haplotypes 1–4. B1–B11: mtDNA Cytb haplotypes 1–11. M1-M21: the combined haplotype 1–21. The same applies below.

Eleven unique mtDNA Cytb haplotypes (B1–B11, NCBI accession nos. PP056522-PP056532, Table S1) were identified from 310 individuals. The haplotype content was 3.5% (11/310). Haplotypes B1, B2, and B3 were present across all the surveyed hosts, representing 58.06%, 16.45%, and 13.55% of the individuals, respectively. Haplotype B4 was observed in nine host populations, accounting for 6.13% of the individuals. Haplotype B8 was found in six populations (3.23% of the individuals) and was consistently present in weed populations across both sampling years. The remaining infrequent haplotypes exhibited irregular distributions among different populations. The mean genetic distance among the Cytb haplotypes was 0.006, ranging from 0.002 to 0.012. In total, 21 combined haplotypes of COI, COII, and Cytb (COM1–COM21) were identified from 295 individuals. The haplotype content was 7.1% (21/295). Haplotypes COM1, COM2, and COM3 were present across all the surveyed hosts, representing 54.24%, 16.61%, and 10.17% of the individuals, respectively. The infrequent haplotypes COM16 and COM17 were detected in the soybean and weed populations. The mean genetic distance among the COM haplotypes was 0.002, ranging from 0.001 to 0.004.

Cluster analysis of all the haplotypes did not reveal a clear host pattern (Figure 4). Many nodes were supported by low bootstrap confidence levels. All the haplotypes were distinct from the outgroup species. Haplotypes with shared variable sites formed strongly supported clades (e.g., the clade comprising COM2 and COM14 as well as the clade comprising COM5 and COM15). A clade containing haplotypes C2 and C7, each with a single variable site, exhibited a broader distribution. Infrequent haplotypes B5, B6, and B11 formed sister clades, and the three haplotypes had unique single variable sites. In a clade containing haplotypes B2 and B4, the two more widely distributed haplotypes showed similarities in variable sites. These results indicated a shared evolutionary and distributional pattern.

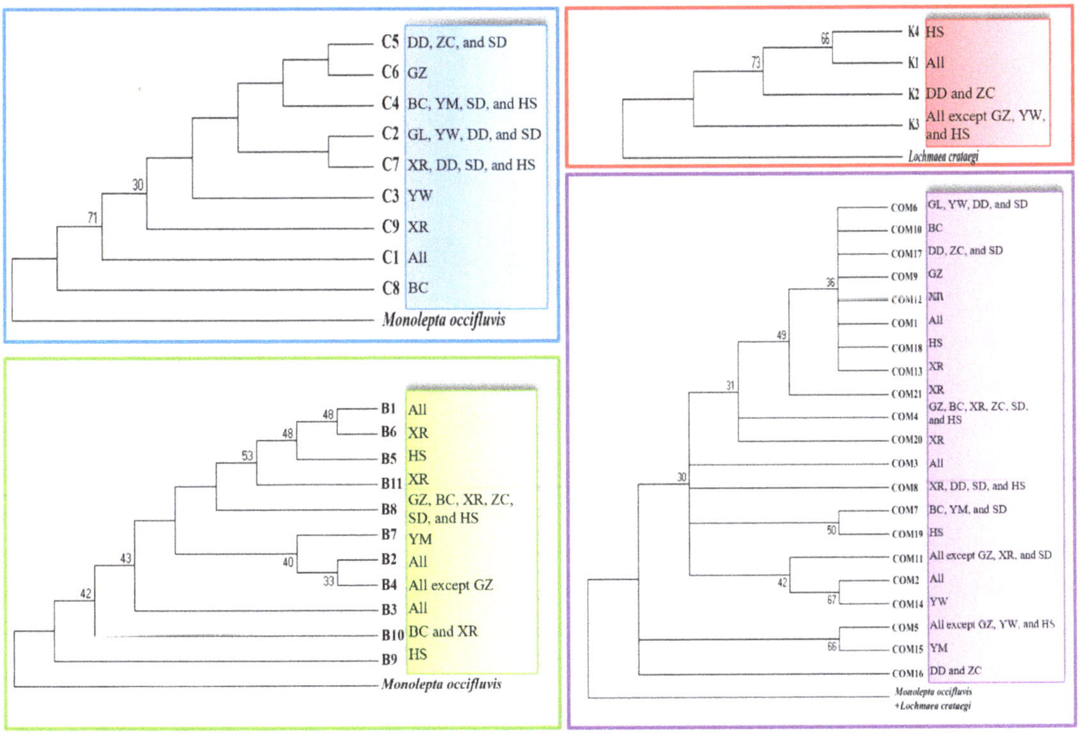

Figure 4. The phylogenetic relationships (NJ analysis) of leaf beetles were examined based on COI, COII, Cytb, and the combined haplotypes. The outgroup taxa were *Monolepta occifluvis* Gressitt and Kimoto (Coleoptera: Chrysomelidae) (Sequence ID: NC_045838.1) and *Lochmaea crataegi* (Forster) (Coleoptera: Chrysomelidae) (Sequence ID: OX387429.1). Bootstrap values were generated from 1000 replicates, and values <30% are not shown. COM1-COM21: the combined haplotype 1–21.

3.4. Genetic Diversity and AMOVA

The genetic diversity indices are shown in Table 3. Overall, the *Hd*, *K*, and *Pi* values of all the COI samples were 0.1412, 0.1456, and 0.0002, respectively. The haplotype range was 2–5, with a mean value of 2.9. The samples from the rice population had the most haplotypes. The mean *Hd* was 0.1641 (range: 0.0555–0.4727). The maize (L) population had the highest *Hd*, whereas the weed population had the lowest. Among the host populations, the average *K* value was 0.1703, ranging from 0.0555 to 0.5090. The *Pi* values based on host populations varied 0–0.0008, with an average value of 0.0002. The *Gst*, *Fst*, and *Nm* values were 0.0019, 0.0138, and 11.88, respectively. For the COII samples, the overall *Hd*, *K*, and *Pi* values were 0.0893, 0.1663, and 0.0003, respectively. The haplotype range was 1–3, with a mean value of 2. Samples from the soybean and weed populations had the most haplotypes, whereas those from the maize (L) and millet populations had the lowest. The mean *Hd* was 0.0827 (range: 0–0.1921). The sorghum populations had the highest *Hd*. Among the host populations, the average *K* value was 0.1552 (range: 0–0.3842). The *Pi* values based on the host populations were 0–0.0008, with an average value of 0.0003. The *Gst*, *Fst*, and *Nm* values were −0.0019, 0.0006, and 20.73, respectively.

For the Cytb samples, the overall *Hd*, *K*, and *Pi* values were 0.6144, 1.6229, and 0.0037, respectively. The haplotype range was 4–8, with a mean value of 5.2. Samples from the sunflower population had the most haplotypes. The mean *Hd* was 0.6026 (range: 0.4137–0.7468). The peanut population had the highest *Hd*, whereas the sorghum population had the lowest. Among the host populations, the average *K* value was 1.6109, with a range of 1.2315–1.8662. The *Pi* values based on the host populations varied 0.0028–0.0043, with an average value of 0.0037. The *Gst*, *Fst*, and *Nm* values were 0.013, 0.0006, and 13.56, respectively. For the combined fragment, the overall *Hd*, *K*, and *Pi* values were 0.6663, 1.9295, and 0.0013, respectively. The haplotype range was 5–10, with a mean value of 7.6. The mean *Hd* was 0.6717 (range: 0.4746–0.8). Among the host populations, the average *K* value was 1.9408, ranging from 1.4492 to 2.2909. The *Pi* values based on host populations varied 0.0009–0.0015, with an average value of 0.0013. The *Gst*, *Fst*, and *Nm* values were 0.008, −0.0019, and 14.2, respectively.

The AMOVA revealed that most of the total variation was within populations (Table 4). No clear pattern emerged based on genetic distance among the host populations (Table 5). Stable genetic distances were noted when comparing sorghum with the cabbage, sunflower, and peanut populations, and when comparing weed with the peanut, cabbage, sunflower, and maize populations. There was also stability when comparing maize with the millet, sorghum, sunflower, and soybean populations. The genetic distance between maize (L) and the other host populations for COI was found to be relatively large. This may be attributed to the small sample size of maize (L) populations. Additionally, differences in haplotype distribution among maize (Z) populations may contribute to this pattern. A similar phenomenon was observed in the millet and sorghum populations for COII and the peanut and sorghum populations for Cytb. No genetic distance was found between the soybean and weed populations.

Table 3. Leaf beetle genetic diversity among host populations.

Population Code	Number of Haplotypes (*H*)				Haplotype Diversity (*Hd*)				Average Number of Nucleotide Differences (*K*)				Nucleotide Diversity (*Pi*)			
	COI	COII	Cytb	COM	COI	COII	Cytb	COM	COI	COII	Cytb	COM	COI	COII	Cytb	COM
GZ	2	1	4	5	0.0800	0.0000	0.5158	0.6100	0.0800	0.0000	1.3836	1.5733	0.0001	0.0000	0.0032	0.0010
BC	3	2	6	8	0.1354	0.0665	0.6781	0.6970	0.1379	0.1333	1.7770	2.0098	0.0002	0.0003	0.0041	0.0013
GL	2	2	4	6	0.1330	0.1921	0.4137	0.5197	0.1330	0.3842	1.2315	1.7487	0.0002	0.0008	0.0028	0.0011
XR	3	2	8	10	0.1354	0.0666	0.7333	0.7635	0.1379	0.1333	1.6689	1.9064	0.0002	0.0003	0.0038	0.0012
YW	3	1	4	6	0.4727	0.0000	0.5606	0.8000	0.5090	0.0000	1.6818	2.2909	0.0008	0.0000	0.0039	0.0015
YM	2	2	5	7	0.1024	0.1423	0.5882	0.6899	0.1024	0.2846	1.8461	2.1479	0.0001	0.0006	0.0042	0.0014
DD	4	3	4	9	0.1536	0.1484	0.5395	0.6201	0.1578	0.2456	1.6134	2.0571	0.0002	0.0005	0.0037	0.0013
ZC	2	3	5	8	0.0555	0.0987	0.4628	0.4746	0.0555	0.1474	1.2871	1.4492	0.0000	0.0003	0.0029	0.0009
SD	5	2	5	9	0.2610	0.0666	0.6871	0.7758	0.2758	0.1333	1.7536	2.1674	0.0004	0.0003	0.0040	0.0014
HS	3	2	7	8	0.1126	0.0454	0.7468	0.7664	0.1142	0.0909	1.8662	2.0570	0.0001	0.0002	0.0043	0.0013
Total	9	4	11	21	0.1412	0.0893	0.6144	0.6663	0.1456	0.1663	1.6229	1.9295	0.0002	0.0003	0.0037	0.0013

Table 4. Information regarding the AMOVA.

Source of Variation	Variance Components				Percentage of Variation			
	COI	COII	Cytb	COM	COI	COII	Cytb	COM
Among populations	0.00085	−0.00041	0.006	0.00377	1.17067	−0.4957	0.73946	0.39074
Within populations	0.07205	0.08352	0.80607	0.96140	98.82933	100.4957	99.26054	99.60926

Table 5. Genetic distances among leaf beetle host populations were analyzed using molecular markers: COI (below the diagonal, upper values), Cytb (below the diagonal, lower values), COII (above the diagonal, upper values), and the combined fragment (above the diagonal, lower values).

	GZ	BC	GL	XR	YW	YM	DD	ZC	SD	HS
GZ		0.00000 0.00000	0.07949 0.00000	0.00000 0.00000	0.00000 0.00000	0.03975 0.01019	0.03187 0.00000	0.00779 0.00000	0.00000 0.00000	0.00000 0.01960
BC	0.00000 0.00000		0.00441 0.01550	0.00000 0.00000	0.00000 0.00000	0.00000 0.00000	0.00000 0.00000	0.00000 0.01543	0.00000 0.00000	0.00000 0.00000
GL	0.02132 0.00000	0.01818 0.01981		0.00441 0.01496	0.01868 0.00000	0.00000 0.05745	0.00000 0.00000	0.00683 0.00000	0.00441 0.05175	0.05724 0.08945
XR	0.00000 0.00000	0.00000 0.00000	0.01818 0.02218		0.00000 0.00000	0.00000 0.00000	0.00000 0.00000	0.00000 0.01217	0.00000 0.00000	0.00000 0.00186
YW	0.14270 0.00000	0.12279 0.00000	0.02783 0.00000	0.12279 0.00000		0.00000 0.00000	0.00000 0.00000	0.00000 0.00000	0.00000 0.00000	0.00000 0.00000
YM	0.01353 0.04262	0.00000 0.00000	0.03471 0.08112	0.01333 0.01104	0.17475 0.00000		0.00000 0.00000	0.00000 0.06583	0.00000 0.00000	0.02462 0.00000
DD	0.00000 0.00000	0.00000 0.00000	0.00000 0.00000	0.00000 0.00000	0.08682 0.00000	0.01093 0.00690		0.00000 0.00000	0.00000 0.00000	0.01303 0.01555
ZC	0.00232 0.00000	0.00304 0.01335	0.03031 0.00000	0.00304 0.01509	0.20990 0.00000	0.01743 0.07804	0.00000 0.00000		0.00000 0.05830	0.00000 0.09312
SD	0.00000 0.02277	0.00000 0.00000	0.00000 0.06908	0.00000 0.00000	0.03676 0.00000	0.00000 0.00000	0.00000 0.00000	0.00000 0.06218		0.00000 0.00000
HS	0.00000 0.04104	0.00000 0.00000	0.02053 0.08785	0.00000 0.00000	0.15081 0.00000	0.00000 0.00000	0.00000 0.01361	0.00028 0.08216	0.00000 0.00000	

4. Discussion

The leaf beetle *M. hieroglyphica* was present from July to October in 2022 and 2023, with daily catch numbers peaking from late July to mid-August over two successive years. This period coincided with the crucial growing season in Northeast China, consistent with previous reports [7]. The pest appeared earlier in maize, soybean, and weed hosts, persisting into later periods in maize, maize (L), and cabbage fields. No beetles were observed after crop harvest, suggesting a close relationship between their occurrence and plant growth. This result appears to be associated with the host transfer migration of some *M. hieroglyphica* individuals. Based on these observations, we propose a possible migration pathway: the leaf beetle initially appears in soybean and weed fields in early July, subsequently disperses to other crops, and eventually settles in cabbage and maize (L) fields by the later periods. The spatial dynamics and host transfer migration pathway align with previous reports [5]. Considering the insect's lifespan, most host transfers may occur within a single step.

In terms of occurrence, the leaf beetle showed high numbers in maize and soybean fields. Given that maize and soybean are dominant crops in the region, and the pest's occurrence area continues to expand [3], it is conceivable that its increasing numbers may result in considerable economic losses in the future. The highest number of beetles was found in the maize (L) field, indicating the insect's preference for young leaves [6]. This finding mirrors observations in large farmland areas, where late-seeded maize is often severely damaged by the leaf beetle. A certain number of leaf beetles occurred in weed, sorghum, millet, and peanut fields, suggesting that these plants are suitable hosts. The low occurrence in sunflower fields could be explained by the widely spaced planting, whereas almost no individuals were observed in the wheat field, possibly due to earlier harvest times. Thus, it appears that wheat in the region is not affected by leaf beetles.

Li et al. [13] suggested that the genetic diversity of geographic populations in southern China was higher than that in northern China, possibly due to higher temperatures and more generations. Our research aimed to determine the level of genetic diversity among host populations. Genetic diversity is considered an important indicator of a species' adaptive capacity in different environments [29–38]. Species with higher genetic diversity are expected to exhibit local adaptation and greater individual numbers [30,39]. However, there was no evidence to support a consistent relationship between genetic diversity and occurrence in the present study. For instance, high levels of genetic diversity were detected in the populations of maize, soybean, cabbage, and sunflower, but occurrence levels were not consistent among these hosts.

Different mtDNA fragments may evolve at different rates. More variable sites and haplotypes were found in the COI and Cytb fragments than in the COII fragment. Combining multiple mtDNA fragments can greatly increase the value of research [40–42]. Notably, the haplotype contents of *M. hieroglyphica* geographic populations previously reported in northern and southern China (2.9% and 5.7%, respectively) were higher than those in our study (1.2%) based on the same COII fragment [13,14]. It appears that the host populations exhibited a lower degree of haplotype content compared with geographic populations. Various haplotypes, including C1, K1, B1, B2, and B3, were found in all the host populations. These are ancestral haplotypes, the most frequent and widespread haplotype, which show robust adaptation to the local environment [39,40,43].

Based on the haplotype network and NJ tree, no distinct host pattern was formed. The haplotypes exhibited a weak host correlation, which was supported by other analyses. Estimates of the overall genetic differentiation coefficient (*Gst* and *Fst*) were low, and all *Nm* values exceeded 11, with $Nm > 4$ indicating strong gene flow in the analyzed populations [44,45]. The high level of gene flow was likely due to host transfer migration, which can prevent genetic divergence among host populations. Combined with the AMOVA, this analysis showed that most genetic variation was within populations. These findings do not strongly support the formation of host races. In contrast to previous studies on geographic populations, populations in northern and southern China exhibited similar levels of genetic divergence, with limited gene flow observed and some degree of variation among populations [13,14]. Therefore, based on the same COII fragment, host populations exhibited a higher degree of gene flow compared with geographic populations. This may be attributed to the limited flight capability of the studied species [6]; leaf beetles are not considered migratory, being capable of dispersal over short distances only.

Geographic isolation over extended periods can lead to genetic differentiation and the emergence of new subspecies [46–49]. Additionally, host specialization has been known to contribute to the formation of host races [50–53]. Although clear evidence for host-adapted races of the leaf beetle was not found in the current study, the data on haplotype distribution and genetic distances support some level of host genetic divergence. The evidence from haplotype distribution includes the following: (i) infrequent haplotypes, especially single haplotypes widespread among hosts; (ii) the consistent presence of infrequent haplotypes in the same hosts (C4 in maize and B8 in weeds) across both sampling years; and (iii) early occurrence of *M. hieroglyphica* populations in soybean and weed hosts, corresponding to the presence of specific infrequent haplotypes (e.g., C5 and K2) in these populations. These molecular data align with prior field observations. There was no evidence supporting genetic distance between the soybean and weed populations, although some degree of distance existed between the soybean and cabbage populations, likely due to differences in the occurrence period. The leaf beetle initially infests soybean and maize fields before appearing in cabbage and maize (L) fields, suggesting variance in genetic backgrounds associated with occurrence periods. Furthermore, results of genetic variation analysis coincided with the species' morphological and biological characteristics. The peak periods of the leaf beetle differed among the host plant species, and there was a close correlation between emergence periods and body types [7]. Differences in life history and fecundity among host plants have also been reported [15]. Host divergence may contribute to these differences.

In recent years, leaf beetle damage to soybean, maize, and other crops has markedly increased in Northeast China. Our data on occurrence periods, occurrence levels, and genetic structures across the major crops in the region are crucial for developing effective pest control strategies. Although this study presents initial findings, its primary limitations include a relatively small genotyping sample size and a narrow geographical scope. To achieve more comprehensive results, further research should incorporate larger sample sizes and broader geographical ranges.

Supplementary Materials: The following supporting information can be downloaded at: https://www.mdpi.com/article/10.3390/insects16060605/s1, Table S1: Haplotype information.

Author Contributions: Conceptualization, W.S. and Y.G.; methodology, W.S.; software, X.Z.; validation, Y.G.; formal analysis, W.S.; investigation, J.Z.; resources, Y.G.; data curation, X.Z.; writing—original draft preparation, W.S.; writing—review and editing, Y.G.; visualization, X.Z.; supervision, Y.G.; project administration, J.Z.; funding acquisition, W.S. All authors have read and agreed to the published version of the manuscript.

Funding: This research was funded by the Jilin Province Science and Technology Development Plan (20240101220JC); Chinese Academy of Sciences Guide Plan (XDA28080204); and the National Classification of Project (2023YFD15011027).

Data Availability Statement: The data presented in this study are contained within the article and Supplementary Materials. All of the sequence data were deposited in the NCBI Genbank database under accession numbers PP038011–PP038019 and PP056518–PP056532.

Conflicts of Interest: The authors declare no conflicts of interest.

References

1. Sun, W.; Hu, G.; Su, Q.; Wang, Y.; Yang, W.; Zhou, J.; Gao, Y. Population source of third-generation oriental armyworm in Jilin, China, determined by entomology radar, trajectory analysis, and mitochondrial COI sequences. *Environ. Entomol.* **2022**, *51*, 621–632. [CrossRef] [PubMed]
2. Sun, W.; Su, Q.; Yang, W.; Zhou, J.; Gao, Y. Genetic diversity and gene flow observed in two cereal aphid (Hemiptera: Aphididae) species and populations in the Chinese corn belt region. *J. Entomol. Sci.* **2022**, *57*, 363–379. [CrossRef]
3. He, Q.; Song, X.; Ma, H.; Yin, Y. The complete mitochondrial genome of *Monolepta hieroglyphica* (Motschulsky) (Coleoptera: Chrysomelidae). *Mitochondrial DNA Part B* **2021**, *6*, 2363–2365. [CrossRef]
4. Zheng, F.; Jiang, H.; Jia, J.; Wang, R.; Zhang, Z.; Xu, H. Effect of dimethoate in controlling *Monolepta hieroglyphica* (Motschulsky) and its distribution in maize by drip irrigation. *Pest. Manag. Sci.* **2019**, *76*, 1523–1530. [CrossRef]
5. Chinese academy of sciences, Institute of zoology. Research brief on *Monolepta hieroglyphica* (Motschulsky). *Acta Entomol. Sin.* **1979**, *22*, 115–117, (In Chinese with English abstract)
6. Chen, G.; Yin, W.; Li, Q.; Hu, H. Research progress on *Monolepta hieroglyphica* (Motschulsky). *China. Plant. Prot.* **2016**, *36*, 19–26, (In Chinese with English abstract)
7. Zhao, X.; Zheng, X.; Guo, J.; Liu, Y.; Luo, B.; Wang, L.; Wang, L.; Liu, Y.; Li, Q.; Wang, Z. Occurrence of *Monolepta hieroglyphica* adults in corn fields in Qiqihar. *Chin. J. Appl. Entomol.* **2021**, *58*, 979–984, (In Chinese with English abstract)
8. Gao, S.; Meng, W.; Zhang, L.; Yue, Q.; Zheng, X.; Xu, J. *Parametarhizium* (*Clavicipitaceae*) gen. nov. with two new species as a potential biocontrol agent isolated from forest litters in Northeast China. *Front. Microbiol.* **2021**, *12*, 627744. [CrossRef]
9. Yan, J.; Tan, Y.; Lv, F.; Wang, Y.; Zhang, Y.; Chi, D. Effect of jasmonate treatments on leaves of Rosa rugosa 'Plena' and detoxification enzymes and feeding of adult *Monolepta hieroglyphica*. *J. For. Res.* **2021**, *32*, 1253–1261. [CrossRef]
10. Zhang, X.; Zhang, R.; Li, L.; Yang, Y.; Ding, Y.; Guan, H.; Wang, X.; Zhang, A.; Wen, H. Negligible transcriptome and metabolome alterations in RNAi insecticidal maize against *Monolepta hieroglyphica*. *Plant. Cell. Rep.* **2020**, *39*, 1539–1547. [CrossRef]
11. Ge, Y.; Shi, C.; Bai, M.; Cao, Z.; Cao, L.; Wang, Z.; Dong, J.; Wang, Y. Molecular data confirm *Monolepta hieroglyphica* (Motschulsky, 1858) and *M. quadriguttata* (Motschulsky, 1860) being synonyms of *M. signata* (Oliver, 1808). *Insect. Syst. Evol.* **2023**, *54*, 402–412. [CrossRef]
12. Li, W.; Shen, S.; Chen, H. Mitochondrial genome of *Monolepta hieroglyphica* (Coleoptera: Chrysomeloidea: Chrysomelidae) and phylogenetic analysis. *Mitochondrial DNA Part B* **2021**, *6*, 1541–1543. [CrossRef]
13. Liang, R. Genetic differentiation of *Monolepta hieroglyphica* (Motschulsky) (Coleoptera: Chrysomelidae) from North China Based on mtDNA COII and Cytb and *Wolbachia* Infection Detection. Master's Thesis, Shenyang Agricultural University, Shenyang, China, 2011. (In Chinese with English abstract)
14. Li, J.; Zhang, X.; Xu, L.; Shen, Y.; Li, X.; Wang, Z. Genetic structure and Wolbachia infection in geographical populations of *Monolepta hieroglyphica* (Coleoptera: Chrysomelidae) in South China. *Acta Entomol. Sin.* **2021**, *64*, 730–742, (In Chinese with English abstract)
15. Zhang, M.; Cui, J.; Xu, W.; Qin, H.; Liu, D.; Wang, X.; Shi, S. Influences of several major crop hosts on adult oviposition of *Monolepta hieroglyphica* (Motschulsky). *Chin. Agric. Sci. Bull.* **2015**, *31*, 81–84, (In Chinese with English abstract)
16. O'Neill, K.; Larson, D.; Kemp, W. Sweep sampling technique affects estimates of the relative abundance and community composition of grasshoppers (Orthoptera: Acrididae). *J. Agr. Urban Entomol.* **2003**, *19*, 125–131.
17. Whipple, S.; Brust, M.; Hoback, W.; Farnsworth-Hoback, K. Sweep sampling capture rates for rangeland grasshoppers (Orthoptera: Acrididae) vary during morning hours. *J. Orthoptera Res.* **2010**, *19*, 75–80. [CrossRef]
18. Avise, J. Ten unorthodox perspectives on evolution prompted by comparative population genetic findings on mitochondrial DNA. *Annu. Rev. Genet.* **1991**, *25*, 45–69. [CrossRef] [PubMed]
19. Harrison, H. Animal mtDNA as a genetic marker in population and evolutionary biology. *Trends. Ecol. Evol.* **1989**, *4*, 6–11. [CrossRef]
20. Moritz, C.; Dowling, T.; Brown, W. Evolution of animal mitochondrial DNA: Relevance for population biology and systematics. *Ann. Rev. Ecol. Evol. Syst.* **1987**, *18*, 269–292. [CrossRef]
21. Rodrigues, B.; Costa, G.; Godoy, R.; Júnior, A.; Cella, W.; Ferreira, G.; Medeiros, G.; Shimabukuro, P. Molecular and morphometric study of Brazilian populations of *Psychodopygus davisi*. *Med. Vet. Entomol.* **2023**, *38*, 83–98. [CrossRef]
22. Sandra, D.; Luis, M.; Marcelo, B.; Octavio, M.; Juan, M.; Santiago, N.; Matias, S.; Evelina, T.; José, M.; José, D.; et al. Low genetic diversity of the only clade of the tick *Rhipicephalus microplus* in the Neotropics. *Pathogens* **2023**, *12*, 1344. [CrossRef] [PubMed]

23. Sherzada, S.; Hussain, N.; Hussain, A.; Mohame, A.; Khan, S. Diversity and genetic structure of freshwater shark *Wallago attu*: An emerging species of commercial interest. *Environ. Sci. Pollut. Res.* **2024**, *31*, 15571–15579. [CrossRef] [PubMed]
24. Wang, X.; Wang, Z. *The Atlas of Maize Diseases, Insects, and Grass Pests in China*; Agriculture Press: Beijing, China, 2018; pp. 176–178. (In Chinese)
25. Tamura, K.; Dudley, J.; Nei, M.; Kumar, S. MEGA4: Molecular evolutionary genetics analysis (MEGA) software version 4.0. *Mol. Biol. Evol.* **2007**, *24*, 1596–1599. [CrossRef] [PubMed]
26. Librado, P.; Rozas, J. DnaSP V5: A software for comprehensive analysis of DNA polymorphism data. *Bioinformatics* **2009**, *25*, 1451–1452. [CrossRef] [PubMed]
27. Bandelt, H.; Forster, P.; Rohl, A. Median-joining networks for inferring intraspecific phylogenies. *Mol. Biol. Evol.* **1999**, *16*, 37–48. [CrossRef]
28. Excoffier, L.; Laval, S.; Schneider, S. Arlequin (version 3.0): An integrated software package for population genetics data analysis. *Evol. Bioinform. Online* **2007**, *23*, 47–50. [CrossRef]
29. Gedifew, S.; Demelasa, H.; Abate, A.; Abebe, T. Association of quantitative traits and genetic diversity in Ethiopian sesame (*Sesamum indicum* L.) genotypes. *Heliyon* **2024**, *10*, e26676. [CrossRef]
30. Ramiro, M.; Asier, G.; Farnando, A.; Todd, J.; Lynda, A.; Kruger, T.; Mike, J.; Alex, G.; Chris, S.; James, R. Population genetic structure and predominance of cyclical parthenogenesis in the bird cherry-oat aphid *Rhopalosiphum padi* in England. *Evol. Appl.* **2020**, *13*, 1009–1025.
31. Hughes, A.; Inouye, B.; Johnson, M.; Underwood, N.; Vellend, M. Ecological consequences of genetic diversity. *Ecol. Lett.* **2008**, *11*, 609–623. [CrossRef]
32. Khabiya, R.; Choudhary, G.; Sairkar, P.; Silawat, N.; Jnanesha, A.; Kumar, A.; Lal, R. Unraveling genetic diversity analysis of Indian ginseng (*Withania somnifera* (Linn.) Dunal) insight from RAPD and ISSR markers and implications for crop improvement vital for pharmacological and industrial potential. *Ind. Crops Prod.* **2024**, *210*, 118124. [CrossRef]
33. Kononov, A.; Ustyantsev, K.; Wang, B.; Mastro, V.; Fet, V.; Blinov, A.; Yuri, B. Genetic diversity among eight Dendrolimus species in Eurasia (Lepidoptera: Lasiocampidae) inferred from mitochondrial COI and COII, and nuclear ITS2 markers. *BMC Genom. Data* **2016**, *17*, 157. [CrossRef]
34. Misra, V.; Pandey, H.; Srivastava, S.; Sharma, A.; Kumar, R.; Pandey, A.; Singh, S.; Singh, V. Computational analysis of haplotype diversity, phylogenetic variation, and population structure of *Candidatus Phytoplasma aurantifolia* using tuf gene sequences. *Ecol. Genet. Genom.* **2024**, *31*, 100229. [CrossRef]
35. Mooney, K. Genetically based population variation in aphid association with ants and predators. *Arthropod-Plant. Inte.* **2011**, *5*, 1–7. [CrossRef]
36. Andrada-Souza, V.; Silva, J.; Hamada, N. Phylogeography and population diversity of *Simulium hirtipupa* Lutz (Diptera: Simuliidae) based on mitochondrial COI sequences. *PLoS ONE* **2017**, *12*, e0190091. [CrossRef]
37. Wenzel, M.; Webster, L.; Blanco, G.; Burgess, M.; Kerbiriou, C.; Segelbacher, G.; Piertney, S.; Reid, J. Pronouced genetic structure and low genetic diversity in European red billed chough (*Pyrrhocorax pyrrhocorax*) populations. *Conserv. Genet.* **2012**, *13*, 1213–1230. [CrossRef]
38. Wongsa, K.; Duangphakdee, O.; Rattanawannee, A. Genetic structure of the *Aphis craccivora* (Hemiptera: Aphididae) from Thailand inferred from mitochondrial COI gene sequence. *J. Insect. Sci.* **2017**, *17*, 84. [CrossRef]
39. Tir, M.; Tombari, W.; Khawla, T.; Tarek, H.; Abdeljelil, G.; Mhamed, E. Genetic diversity and population structure of *Sepia officinalis* from the Tunisian cost revealed by mitochondrial COI sequences. *Mol. Biol. Rep.* **2015**, *42*, 77–86.
40. Wang, X.; Yang, X.; Lu, B.; Zhou, L.; Wu, K. Genetic variation and phylogeographic structure of the cotton aphid, *Aphis gossypii*, based on mitochondrial DNA and microsatellite markers. *Sci. Rep.* **2017**, *7*, 1920. [CrossRef]
41. Laopichienpong, N.; Muangmai, N.; Supikamolseni, A.; Twilprawat, P.; Chanhome, L.; Suntrarachun, S.; Peyachoknagul, S.; Srikulnath, K. Assessment of snake DNA barcodes based on mitochondrial COI and Cytb genes revealed multiple putative cryptic species in Thailand. *Gene* **2016**, *594*, 238–247. [CrossRef]
42. Rakauskas, R.; Havelka, J.; Bernoticne, R. Mitochondrial (COI) and nuclear (EF-1α) DNA vatiability of *Rhopalosiphum padi* and *Rhopalosiphum nymphaeae* (Hemiptera: Aphididae) in Lithuania. *Biologia* **2014**, *69*, 1730–1741. [CrossRef]
43. Thirumaraiselvi, R.; Thangaraj, M. Genetic diversity analysis of Indian salmon, *Eleutheronema tetradactylum* from South Asian countries based on mitochondrial COI gene sequences. *Not. Sci. Biol.* **2015**, *7*, 417–422. [CrossRef]
44. Millar, C.; Libby, W. *Genetics and Conservation of Rare Plants*; Oxford University Press: New York, NY, USA, 1991; pp. 149–170.
45. Whitlock, M.; Mccauley, D. Indirect measures of gene flow and migration: Fst $\neq 1/(4Nm + 1)$. *Heredity* **1999**, *82*, 117–125. [CrossRef] [PubMed]
46. Criniti, A.; Mazzoni, E.; Pecchioni, N.; Rau, D.; Cassanelli, S.; Bizzaro, D.; Manicardi, G. Genetic variability among different Italian populations of the aphid *Myzus persicae*. *Caryologia* **2006**, *59*, 326–333. [CrossRef]

47. Mitić, Z.; Nikolić, J.; Jušković, M.; Ranđelović, V.; Nikolić, B.; Zlatković, B. Geographic differentiation of *Abies alba*, *A. x borisii-regis*, and *A. cephalonica* populations at the Balkan Peninsula based on needle morpho-anatomy. *Trees-Struct. Funct.* **2023**, *37*, 1465–1481. [CrossRef]
48. Ortego, J.; Bonal, R.; Cordero, P.; Aparicio, J. Phylogeography of the Iberian populations of *Mioscirtus wagneri* (Orthoptera: Acrididae), a specialized grasshopper inhabiting highly fragmented hypersaline environments. *Biol. J. Linn. Soc.* **2009**, *97*, 623–633. [CrossRef]
49. Palombo, N.; Carrizo, G. Geographical patterns of genetic variation in Locoto Chile (*Capsicum pubescens*) in the Americas inferred by genome-wide data analysis. *Plants* **2022**, *11*, 2911. [CrossRef]
50. Anstead, J.; Burd, J.; Shufran, K. Mitochondrial DNA sequence divergence among *Schizaphis graminum* (Hemiptera: Aphididae) clones from cultivated and non-cultivated hosts: Haplotype and host associations. *Bull. Entomol. Res.* **2002**, *92*, 17–24. [CrossRef]
51. Charaabi, K.; Carletto, J.; Chavigny, P.; Marrakchi, M.; Makni, M.; Vanlerberghe-Masutt, F. Clonal diversity of the melon aphid *Aphis gossypii* (Glover) in Tunisia is structured by host plants. *Bull. Entomol. Res.* **2008**, *98*, 333–341. [CrossRef]
52. Corre, V.; Reibel, C.; Kati, V.; Gibot-Leclerc, S. Host-associated genetic differentiation and origin of a recent host shift in the generalist parasitic weed *Phelipanche ramosa*. *Ecol. Evol.* **2023**, *13*, e10529.
53. MacDonald, Z.; Snape, K.; Roe, A.; Sperling, F. Host association, environment, and geography underlie genomic differentiation in a major forest pest. *Evol. Appl.* **2022**, *15*, 1749–1765. [CrossRef]

insects

MDPI

Article

What Is the Relationship Between Efficacy of Seed Treatment with Insecticides Against *Dalbulus maidis* (Delong and Wolcott) (Hemiptera: Cicadellidae) Healthy and Infected with Spiroplasm in the Corn Stunt Control?

Ana Carolina M. Redoan [1], Vinicius M. Marques [2], Poliana S. Pereira [1], Ivênio R. de Oliveira [1], Dagma D. Silva-Araújo [1], Luciano V. Cota [1], Marcos Antonio M. Fadini [2], Charles M. Oliveira [1], Diego D. Rafael [1] and Simone Mendes [1,*]

[1] Brazilian Agricultural Research Corporation (Embrapa), Sete Lagoas 35702-098, Brazil; ac.redoan@gmail.com (A.C.M.R.); ivenio.rubens@embrapa.br (I.R.d.O.); dagma.silva@embrapa.br (D.D.S.-A.); charles.oliveira@embrapa.br (C.M.O.)

[2] Department of Biosystems Engineering, Federal University of São João del Rei (UFSJ), São João del Rei 36307-352, Brazil

* Correspondence: simone.mendes@embrapa.br; Tel.: +55-31-3027-1136/+55-31-9216-1864

Simple Summary

The corn leafhopper *Dalbulus maidis* is an insect vector that inoculates phytopathogens into the phloem, compromising nutrition, growth, and corn development. Symptoms of corn stunt include chlorosis, shortened internodes, plant malformation, ear proliferation, and even plant mortality. In more severe cases, it can reduce corn yield by up to 100%. We evaluated the effect of the main insecticides for seed treatment on the control of infective and non-infective leafhoppers, the persistence of the treatment's effect, and its relationship to the expression of disease symptoms and yield. We observed that seed treatment was effective until the V2 corn stage and that infective leafhoppers were more sensitive to insecticides compared to non-infective ones. Plants were more susceptible to the disease in the early vegetative stages, significantly reducing corn yield.

Abstract

Seed treatments with insecticides are important tools for managing corn stunting disease complex (CSDC) transmitted by *Dalbulus maidis* (Hemiptera: Cicadellidae) by reducing the initial leafhoppers' population and, consequently, the risk of pathogen transmission. We evaluated the effect of insecticides used in seed treatment on both healthy and spiroplasma-infected leafhoppers, the persistence of the seed treatment effect on disease symptom severity, and its impact on corn productivity. At the V2 stage, imidacloprid/thiodicarb was the most effective, resulting in 100% mortality of healthy leafhoppers and 85.7% mortality of infective ones, thus preventing spiroplasma transmission. Thiamethoxam and methomyl + fipronil/thiamethoxam showed a high total mortality after 72 h, but only for the infective leafhoppers, with a total mortality of healthy leafhoppers around 40%, reducing the number of plants with symptoms by 80% and 90%, respectively. Our results prove that there is a difference between the chemical molecules and that the infected leafhoppers are more susceptible. Insecticide seed treatment was effective until the V2 growth stage, and imidacloprid/thiodicarb was the most effective product tested. Infective leafhoppers were more susceptible to insecticide seed treatments, and the infestation by the corn leafhopper carrying spiroplasma in the early stages of plant development heavily reduced corn yield.

Keywords: chemical control; leafhopper; integrated management; spiroplasma; *Zea mays*

1. Introduction

The corn leafhopper *Dalbulus maidis* (Delong and Wolcott) (Hemiptera: Cicadellidae) is the main pest currently affecting corn in Brazil and several countries in Latin America [1,2]. This insect is the vector of the bacteria (Mollicute class), spiroplasma (corn stunt spiroplasma—CSS), phytoplasma (maize bushy stunt phytoplasma—MBSP), and viruses (virus maize rayado fino virus (MRFV) and maize striate mosaic virus (MSMV)) associated with the corn stunting disease complex (CSDC) [3,4]. The *D. maidis* feeds on phloem sap and transmits phytopathogens in a persistent and propagative manner, except for MSMV that transmits in a persistent circulative manner, and these phytopathogens occur within the same geographical distribution [5]. CSDC directly impacts the development of maize plants, potentially causing losses of up to 100% of the yield. To manage the insect vector and related diseases, a set of good agricultural practices, primarily preventive, is recommended to reduce the risk of high CSDC incidence [2,6].

Among the pathogens involved in CSDC, phytoplasma and spiroplasma have been considered the most important in terms of negative impacts on grain production [7]. These pathogens can influence the development of *D. maidis* either positively or negatively, although there is no consensus in the results of studies conducted with these pathogens. For example, while some studies suggest that CSS increases the survival of its vector when exposed to temperatures between 10 and 20 °C [8], but does not affect adult leafhopper survival compared to uninfected ones when reared at 26 °C [5], another study found that both CSS and MBSP enhanced the survival of *D. maidis* at cool (15 °C) and warmer (31 °C) temperature conditions [9]. The effects of pathogens on the development of *D. maidis* s have not yet been fully described, and the lack of this knowledge can directly influence management strategies.

Because of the slow development of these pathogens in the vascular systems of maize plants, symptoms of CSDC are usually observed during the grain production stage. The earlier the plants are infected, the greater the losses are [7]. Therefore, management measures for the insect vector should be implemented in the early stages of the crop, particularly between emergence and the V8 stage [6]. In this context, the use of organosynthetic insecticides for seed treatment can be essential in delaying the transmission of pathogens by *D. maidis*, thereby reducing the pathogen load inoculated by the vector in the plants [10–13]. Chemical control through seed treatment is a common tactic that can provide from 10 to 15 days of protection against *D. maidis* [10,13]. In Brazil, there are 74 commercial insecticides registered for controlling *D. maidis*, consisting of combinations of 13 active ingredients from six chemical groups and only five modes of action [14,15]. These products are neuromuscular disruptors, acting on both cholinergic synapses and ion channels in insect neuron axons [16]. However, there are no studies that have demonstrated the efficacy of a seed treatment to control *D. maidis* and to reduce the damage caused by stunting

D. maidis has caused significant economic losses in maize-producing regions, particularly in Argentina, Brazil, and Mexico. In Brazil, the first significative outbreak led to an 85% increase in insecticide usage. In Argentina and Mexico, yield reductions range from 30% to 50% [1,6]. Given that the presence of pathogens (phytoplasma and spiroplasma) can influence the characteristics and behavior of *D. maidis*, it is possible that biotic and abiotic factors, including the application of chemical insecticides, may produce varying results based on the infection status of the insects. The interaction of these chemical products with both healthy and infected insects has not been explored, and understanding this relationship would contribute to a deeper knowledge of this pathosystem. This study aimed to evaluate the effect of insecticide treatment of maize seeds on the mortality of both infected with spiroplasma and healthy adult *D. maidis*, and to investigate how this treatment influences the expression of stunting symptoms and its relationship with reductions in productivity.

2. Materials and Methods

2.1. Insects

The experiment was conducted at the Embrapa Maize and Sorghum Research Center in Sete Lagoas, Minas Gerais, Brazil. We used the method described by Oliveira and Sabato (2017) [7] to obtain the infectious leafhoppers. For five consecutive weeks, 50-day-old maize plants (*Zea mays* L., cultivar LP2020) infected with spiroplasma, grown in pots with 5 kg of soil, were used as the source of infection. Approximately 500 healthy adult *D. maidis* were placed on each infected plant, enclosed in a voile bag tied to the plant stem. After five days in maize source plants (CSS), the leafhoppers were transferred to cages with healthy maize seedlings. After 23 days (latent period), the infective *D. maidis* adults were used to infest the maize plants in the experiments as follows.

2.2. Maize Seeds, Insecticides, and Experiments

The maize hybrid provided by Santa Helena Sementes in Cruz Alta, Brazil (SHS7970 PRO3) was used in the experiments. The insecticides, their active ingredients and application rates are shown in Table 1. The experiments were conducted from September to December 2023 in Sete Lagoas, Minas Gerais, Brazil.

Table 1. Trade name, active ingredient, chemical group, concentration, and dose of the insecticides used in the seed treatment of maize seeds for the control of *Dalbulus maidis* adults in Sete Lagoas/MG, Brazil, between September and December 2023.

Trade Name (1)	Active Ingredient	Chemical Group	Concentration (2)	Dose (3)
Control	Water	----	-----	-----
Nuprid Star FS	Imidacloprid + tiodicarb	Neonicotinoid + carbamate	150 + 450	1.75
Cruiser Opti	Lambda Cyhalothrin + thiamethoxam	Pyrethroid + neonicotinoid	37.5 + 210	1
Ouro fino	Imidacloprid	Neonicotinoid	350	0.08
Poncho	Clothianidin	Neonicotinoid	660	0.1
Cruiser 600 FS	Tiametoxam	Neonicotinoid	600	0.5
Standak Top*+ Impar BR	Thiophanate methyl + fipronil/thiamethoxam	Pyrethroid + neonicotinoid	225 + 250/350	1/0.08

(1) Registered in the Ministry of Agriculture, Livestock and Supply (Mapa) for *Dalbulus maidis*, with the exception of Standak top*. (2) Concentration (g a.i. L^{-1} trade inseticide), (3) dose trade insecticide L 100 Kg^{-1} of seed.

The maize seeds were treated with insecticides and sown on the same day in 200 mL plastic pots filled with fertilized soil, using 3 seeds per pot. After emergence, thinning was performed, leaving one plant per pot. Fertilizers NPK 8-28-16 (Fertilizantes Heringer, Iguatama, Brazil), limestone, and micronutrients (FTE) were used for plant maintenance and growth. Each plant was infested only once, and the experiment was conducted separately for healthy and infective leafhoppers.

In total, there were five infestations (one per time), corresponding to 0, 5, 10, 15, and 20 days after the appearance of the second expanded leaf, which corresponded to the maize's phenological growth stages: V2, V3-V4, V4-V5, V5, and V5-V6. The vegetative stages from V2 to V6 in maize represent key phases in early plant development, each marked by the number of fully visible leaf collars [17]. On each infestation date, seven leafhoppers (infected with CSS or healthy, separately) were placed inside polyethylene terephthalate (PET) bottles, whose upper part was closed with voile fabric, containing a plastic pot with one maize plant. Insect mortality was evaluated for five consecutive days (24, 48, 72, 96, and 120 h). The surviving insects at the end of the fifth day of evaluation were removed and discarded. The design was completely randomized (CRD) with 5 infestation times, 7 insecticide seed treatments (6 insecticides + water), and 2 groups of leafhoppers (infected with CSS and healthy), with 5 replicates corresponded to one plant per pot. Totaling 350 plants, with 175 pots containing plants exposed to healthy leafhoppers and 175 pots containing plants exposed to infective leafhoppers. A total of 2.450 adult *D. maidis* were used (Figure 1).

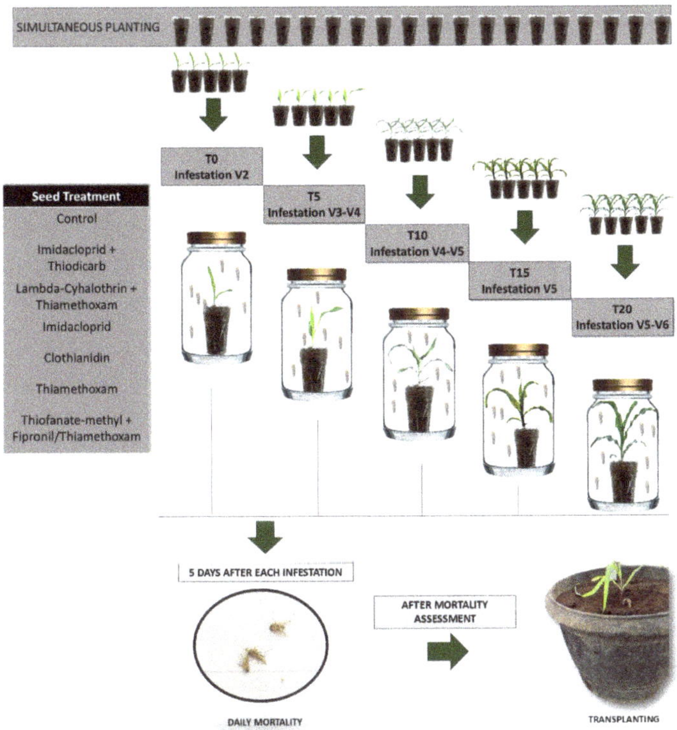

Figure 1. Simplified schematic view of the methodology: sow of treated seeds, leafhopper infestation, mortality assessment, and maize plants exposed to leafhoppers transplanted in the greenhouse.

After the leafhopper mortality evaluation period, maize plants exposed to leafhoppers were transplanted to pots (26 cm diameter and 29 cm high) containing soil fertilized as recommended for maize cultivation. Plants were kept in a greenhouse, where CSS symptoms, plant height (determined from the ground level to the tassel insertion node), and the number and size of ears per plant were measured with a tape measure.

The severity of spiroplasm was assessed in the reproductive maize growth stage (R4), when the consistency of the kernel interior is similar to "dough" [17] We used a symptom rating scale ranging from 1 to 6: 1—no symptoms (health plant), 2—plant with symptoms (reddening or yellowing) with normal height, 3—plant with symptoms and reduction in size,

4—plant with symptoms and early drying, 5—plant with symptoms, reduction in size and early drying, and 6—plant with symptoms severe reduction in size and falling over [18].

After the visual assessments of CSS symptoms, molecular analyses were performed using polymerase chain reaction (PCR) to confirm the infection. Samples were collected 90 days after inoculation, placed in identified falcon tubes and stored in a freezer at −60 °C. DNA extraction for maize plants followed the protocol proposed by Saghai-Maroof et al. (1984) [19] with the modifications suggested by Sousa and Barros (2017) [20]. The PCR test was performed using PCR-based methods as described by Sousa and Barros (2017) [20], using primers CSSF2 (5′-GGC AAA AGA TGT AAC AAA AGT-3′) and CSSR6 (5′-GTT ACT TCA ACA GTA GTT GCG-3′) for the detection of *S. kunkelii* [21].

2.3. Statistical Analysis

For preliminary data analysis, the Shapiro–Wilk (0.05) and Levene (0.05) tests were used to assess normality and homogeneity of variance. The presence of discrepant values (="outliers") was also assessed through direct observation in box-plot graphs. The R statistical environment version 4.4.1 (R Core Team 2024) [22] was used to adjust the models and generate the graphs. The 'MASS', 'car', and 'rstatix' packages were used for the analyses to adjust the models, and the 'ggplot2' package was used to generate the graphs. The numerical database, as well as the scripts for adjusting the models and generating the graphs, were stored in an open GitHub address to increase the checking and reproducibility of the results [23].

Analysis of Mortality of Healthy and Infective Leafhoppers

Generalized Linear Models (GLM) with a binomial distribution were used. This type of distribution is suitable for ratio data that have low variation at extreme values and high variation at intermediate values [23]. Initially, the full model was adjusted, with the explanatory variables being insecticide seed treatment, time of day, and leafhopper infectivity (infective or healthy), as well as the interactions between these variables. The adjusted models were evaluated for overdispersion, the Akaike criterion (AIC), and the random distribution of the residuals. If any terms were found to be insignificant, they were removed, and the models were readjusted. When necessary, the quasibinomial distribution was used to reduce overdispersion [23].

Since daily mortality (%) is a non-normally distributed continuous variable without homogeneity of variance, the Kruskall–Wallis nonparametric test was performed to assess the differences among groups. Data were tested considering different insecticide treatments (imidacloprid/Tiodicarb, λ-cialotrin/Thiametoxam, imidacloprid, Clothinidin, Thiametoxam, and T. methyl + fipronil/Thiametoxam), different times of evaluation (24 h, 48 h, 72 h, 96 h, and 120 h), and if the leafhopper was infective or healthy. The significance threshold value was set at $p < 0.05$. When there were differences, a Mann–Whitney (U test) was performed to evaluate whether the mortality (%) differed between the two groups (post-hoc test).

The study used simple linear regression to analyze the relationship between maize plant height, ear length, and stunting score at a 5% significance level. The Kruskal–Wallis non-parametric test was employed to assess the rank values of stunting scores in relation to the type of insecticide seed treatment, as the score data did not have a defined frequency distribution. Dunn's non-parametric post-hoc test at 5% significance was used to further evaluate the blight score readings taken at the time of planting (5 days after seed treatment).

3. Results

Insecticide seed treatment significantly affected the mortality of *D. maidis* in comparison to the control group (GLM: $\chi^2 = 72.98$, df = 6, $p < 0.001$) (Figure 2A). In addition, there was a significant difference in mortality between healthy and infective leafhoppers, with the spiroplasma-infected leafhoppers showing higher mortality compared to the healthy leafhoppers (GLM: $\chi^2 = 61.98$, df = 6, $p < 0.001$) (Figure 2B).

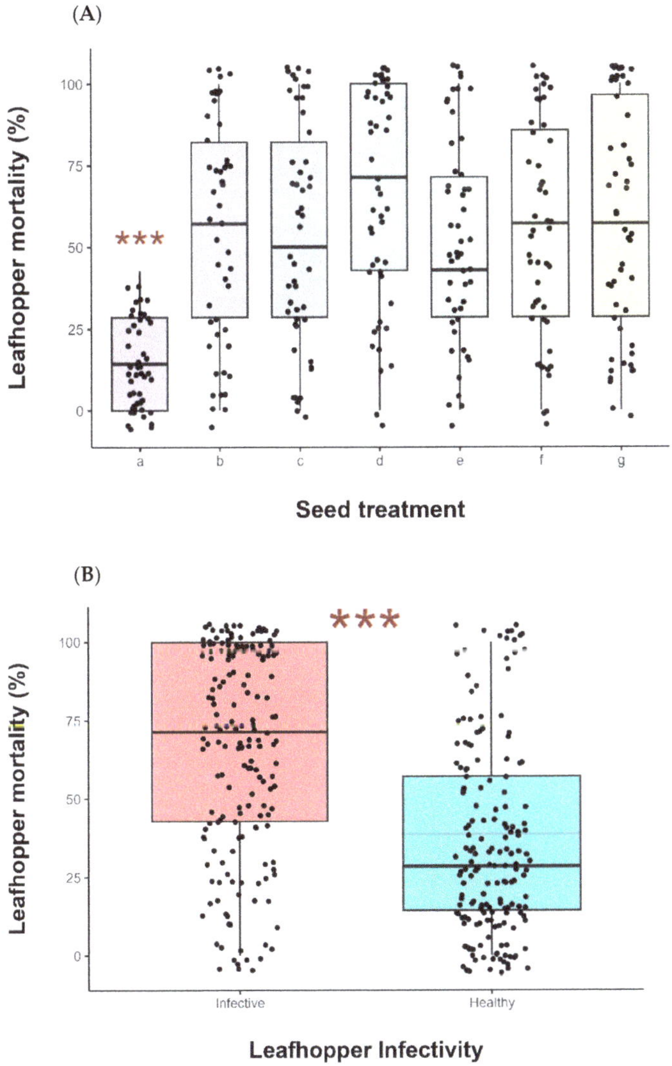

Figure 2. (**A**) Median mortality of all *D. maidis* exposed to insecticides seed treatment. Insecticide seed treatment *x*-axis legend: a = water (=control), b = imidacloprid/thiodicarb, c = Lambda-Cyhalothrin/thiamethoxam, d = imidacloprid, e = Clothianidin, f = Thiamethoxan, and g = thiophanate-methyl + fipronil/thiamethoxam. *** indicates the bar that differs from the others in terms of $p \leq 0.001$. (**B**) Median mortality of infective and healthy *D. maidis* individuals exposed to seed treatment. *** indicates the bar that differs from the others in terms of $p \leq 0.001$.

The interactions among the insecticides seed treatments, leafhopper infectivity, and evaluation time after seed treatment were significant (Table S1). *D. maidis* mortality was affected by individual infectivity ($\chi^2 = 115.19$, df = 1, $p < 0.001$), seed treatment (GLM: $\chi^2 = 93.45$, df = 6, $p < 0.001$), and time after treatment (GLM: $\chi^2 = 90.49$, df = 1, $p < 0.001$). For the control treatment (water), neither time (GLM: $\chi^2 = 2.77$, df = 1, $p = 0.096$) nor leafhopper infectivity (GLM: $\chi^2 = 1.22$, df = 1, $p = 0.268$) affected *D. maidis* mortality (Figure 3A).

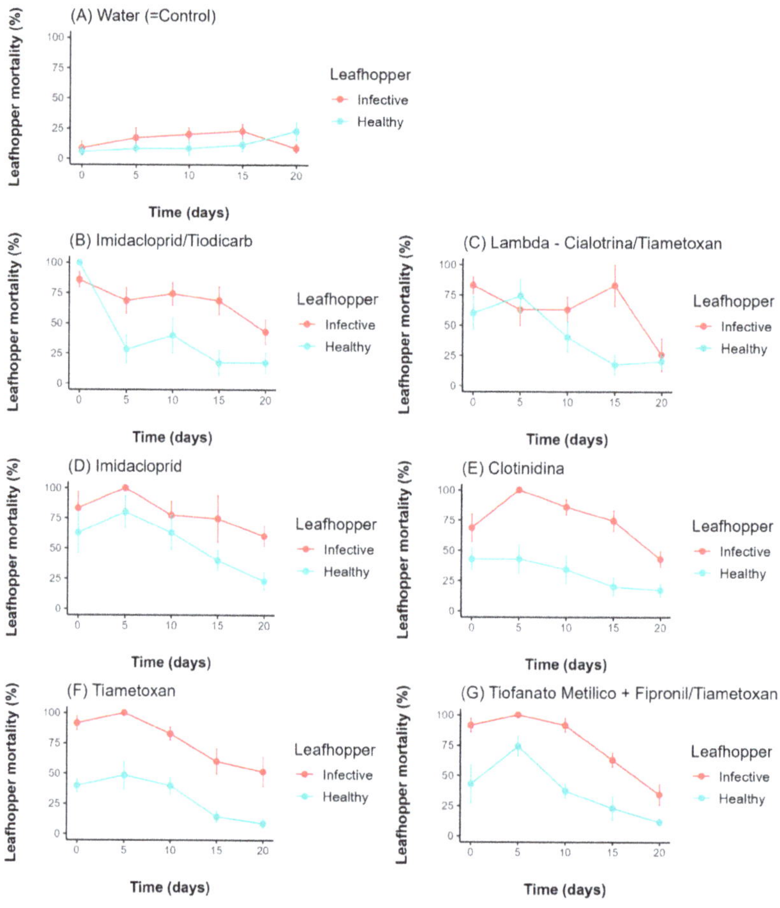

Figure 3. Total mortality (mean ± SEM) of infective and healthy *D. maidis* individuals over time (days) following exposure to control (water) and six insecticide seed treatments. After seed treatment, there were five infestations (one per time), corresponding to 0, 5, 10, 15, and 20 days after the appearance of the second expanded leaf.

For all insecticides used in seed treatments, a higher mortality was observed for both infectious and healthy leafhoppers between time 0 and 5 days after seed treatments. As we can see in the results for seed treatments, both time and infectivity had a significant effect on *D. maidis* mortality, respectively: imidacloprid/thiodicarb ($\chi^2 = 12.89$, df = 1, $p < 0.001$) and ($\chi^2 = 20.80$, df = 1, $p < 0.001$) (Figure 3B); Lambda-Cyhalothrin/thiamethoxam ($\chi^2 = 12.89$, df = 1, $p < 0.001$) and ($\chi^2 = 20.80$, df = 1, $p < 0.001$) (Figure 3C); imidacloprid ($\chi^2 = 9.14$, df = 1, $p = 0.004$) and ($\chi^2 = 9.15$, df = 1, $p = 0.002$) (Figure 3D); Clothianidin ($\chi^2 = 10.94$, df = 1, $p < 0.001$) and ($\chi^2 = 35.59$, df = 1, $p < 0.001$) (Figure 3E); thiamethoxam (χ^2 = 31.26, df = 1, $p < 0.001$) and (χ^2 = 59.36, df = 1, $p < 0.001$) (Figure 3F); and thiophanate-methyl + fipronil/thiamethoxam (χ^2 = 30.21, df = 1, $p < 0.001$) and (χ^2 = 31.21, df = 1, $p < 0.001$) (Figure 3G).

When we evaluated *D. maidis*' daily mortality, the insecticide imidacloprid/thiodicarb showed the highest mortality: 74% within the first 24 h and 100% at 72 h for healthy leafhoppers, and over 80% mortality for infective ones between 24 and 120 h from time zero. This treatment was the most effective in preventing the insect from remaining in contact with the plant, thereby stopping feeding and the transmission of the pathogen (Figure 4).

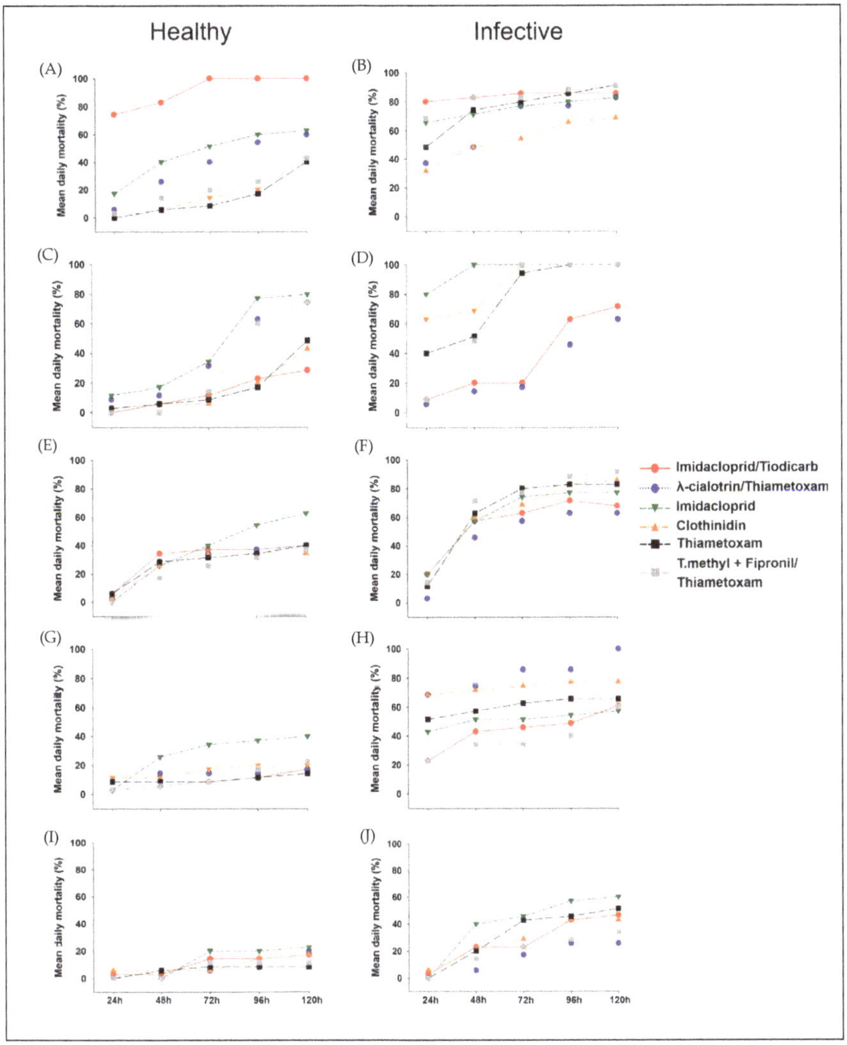

Figure 4. Leafhopper mean daily mortality (%) after different insecticide treatment during five consecutive days (24, 48, 72, 96, and 120 h). Legend: (**A**) = healthy leafhoppers at time 0, (**C**) = healthy leafhoppers at time 5, (**E**) = healthy leafhoppers at time 19, (**G**) = healthy leafhoppers at time 15, and (**I**) = healthy leafhoppers at 20 days after infestation. (**B**) = infective leafhoppers at time 0, (**D**) = infective leafhoppers at time 5, (**F**) = infective leafhoppers at time 10, (**H**) = infective leafhoppers at time 15, and (**J**) infective leafhoppers at 20 days after infestation.

Thiamethoxam and thiophanate-methyl + fipronil/thiamethoxam showed high mortality after 72 h for infective leafhoppers, whereas for healthy leafhoppers, the mortality was below 50%. The seed treatments were efficient, negatively affecting the leafhoppers only at time zero. In other words, the treatment protected the plant for up to 5 days after the emergence of the second fully expanded leaf (V2), or for 11 days after planting (Figure 4).

Visual assessment revealed the impact of leafhoppers infected with S. kunkelli on corn plants (Kruskal–Wallis test: $\chi^2 = 23.08$, df = 6, $p < 0.001$) at time zero, the only case that had statistically significant results (Figure 5). Imidacloprid/thiodicarb treatment demonstrated efficacy in achieving high mortality rates of insects in a short period, which reflected in the lowest stunt score of 1 (no symptoms) (Dunn Test: $p < 0.001$) (Figure 5). Molecular analyses of the plants confined with the infective leafhoppers at time zero confirmed the results of the scores assigned through visual assessment (Figure S1).

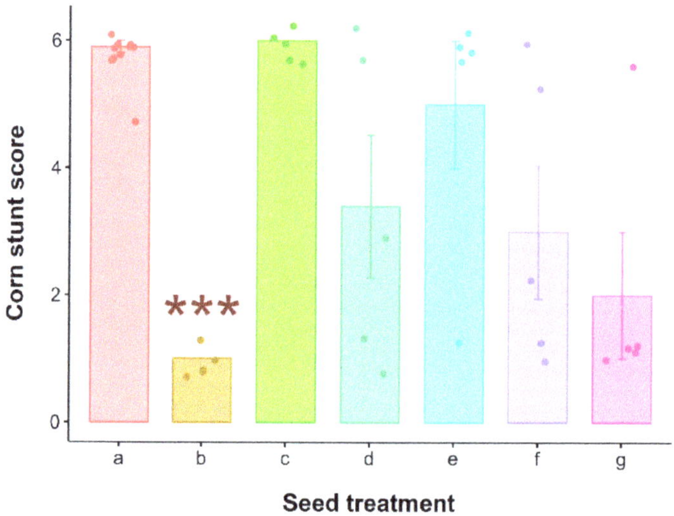

Figure 5. Corn stunt score (mean ± SEM) by seed treatments at time 0 (after 5 days of seed insecticide application). Corn stunt was evaluated at the R4 growth stage using the visual symptom rating scale from 1 to 6 (Silva et al., 2003 [18]). Seed treatment *x*-axis legend: a = water (=control), b = imidacloprid/thiodicarb, c = Lambda-Cyhalothrin/thiamethoxam, d = imidacloprid, e = Clothianidin, f = Thiamethoxan, and g = thiophanate-methyl + fipronil/thiamethoxam. Treatments are represented by different colors. *** indicates the bar that differs from the others in terms of $p \leq 0.001$ by the Dunn a posteriori non-parametric test.

Plants without spiroplasm symptoms had greater vigor and yield than those infested. Both plant height and yield decreased according to disease severity, as indicated by the stunting score (Figure 6). Plants without symptoms (score 1) reached a height of 2.0 m and had ear lengths of 15 cm. In contrast, plants showing symptoms, reduced height, and lodging (score 6) had heights ranging from 1.0 to 1.5 m, with ear lengths below 10 cm.

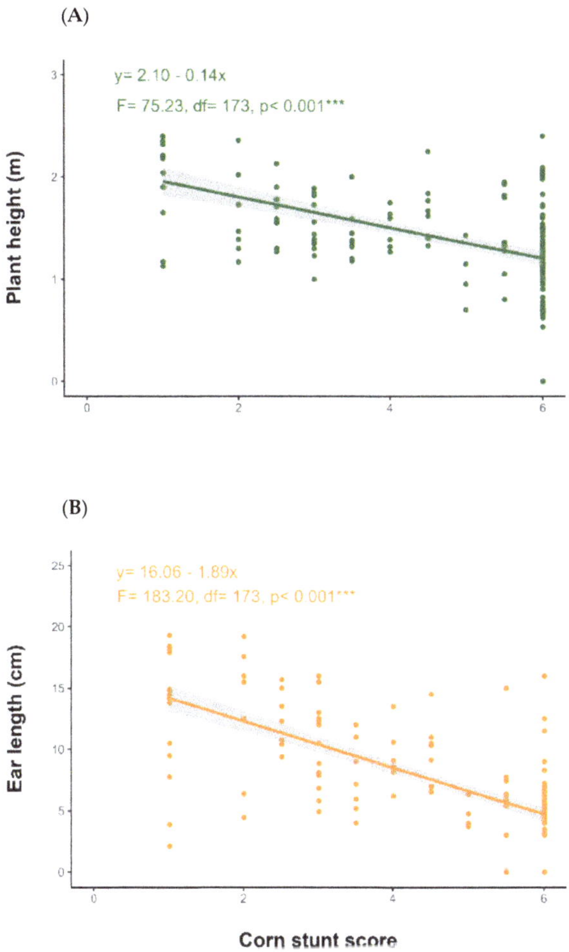

Figure 6. Effect of corn stunt severity on maize height and ear lengths in the greenhouse evaluations. (**A**) Plant height (m) and corn stunt score, (**B**) ear length (m) and corn stunt score. The corn stunt score evaluated using the visual symptom rating scale from 1 to 6 (Silva et al., 2003) [18]. *** indicates the bar that differs from the others in terms of $p \leq 0.001$ by the Dunn a posteriori non-parametric test.

4. Discussion

This result is significant due to the complexity and low efficiency of *D. maidis* and CSDC control in the field. There is a gap in our understanding regarding the bioecology of *D. maidis* and its interactions with the pathogen, resulting in a lack of knowledge about variations in population occurrence in initial growing stages [24] and CSDC transmission. Also, further experiments should be carried out to elucidate questions about the resistance of populations to synthetic organic insecticides.

Given the leafhopper's migratory behavior, it is crucial to synchronize sowing and avoid staggered planting to prevent the presence of corn plants at varying growth stages [25]. Maize crops at different growth stages in nearby areas allow the overlapping of the plant's life cycle. This dynamic favors the multiplication and migration of leafhoppers from mature crops to new ones in the early development stages, efficiently carrying and transmitting mollicutes to these young plants [26].

The use of insecticides has been one of the main strategies for controlling *D. maidis* and CSDC in the field [6,13,27]. Even so, studies published to date on the efficacy of seed treatments for controlling *D. maidis* have prioritized the action of insecticides only on infective leafhoppers [24,27–30]. Thus, the differential in mortality of healthy and infective leafhoppers is very significant, and the contributions of this information to pest management are addressed in the present study. In this context, the results demonstrate that the damage caused by corn stunt disease is not proportional to the size of the vector population but rather to its infectivity rate [2].

Oliveira et al. (2008) [13] verified the efficiency of corn seed treatment for controlling *D. maidis*. In this case, seeds were treated with imidacloprid and thiamethoxam, and healthy leafhoppers were confined with infective ones. These authors found a control efficiency for *D. maidis* adults equal to or greater than 70% up to 30 days after plant emergence, requiring between 4 and 24 h to achieve these mortality rates. Ruegger (2019) [31] evaluated *D. maidis* adult mortality in corn (V3 and V4) via seed treatment with the insecticides chlorantraniliprole, thiamethoxam, and imidacloprid. Only thiamethoxam and imidacloprid demonstrated high efficacy, with mortality rates exceeding 80% on leafhopper populations, reducing their feeding rates. Neves et al. (2021) [28] showed that seed treatment remained effective until the V4 stage when neonicotinoids were used, lowering the disease score (between 2 and 3) for corn stunt and reducing yield losses by 20% to 60%. These results align with those found in this study, where the residual effect of insecticides typically lasted less than three weeks.

Our results showed that infective *D. maidis* adults carrying spiroplasma exhibited higher mortality when exposed to insecticide-treated seeds compared to healthy adults (Figure 2B). This suggests that pathogen infection makes *D. maidis* individuals more susceptible to the action of insecticidal compounds. During the mollicute development, muscle tissues in the body of the leafhopper are damaged [32,33]. As the muscle cells of the midgut and Malpighian tubules are degraded, spiroplasma utilizes nutrients derived from the sarcolemma of the muscle cells. Consequently, the effects of the insecticide are likely enhanced, accelerating the death of the insects, which are physiologically weakened by the pathogen. On each evaluation date, it was observed that the cumulative mortality of the leafhoppers increased over the days following the infestation of the plants. However, when comparing the evaluation dates (0, 5, 10, 15, and 20 days), a reduction in mortality was noted as the leafhoppers were exposed to older plants, indicating a decline over time after the seed treatment (Figure 3). This may be related to the high solubility of neonicotinoids, which leads to the rapid absorption and metabolism of these compounds by the seed after planting. In other words, lower concentrations in the new tissues of the plant reduce the product's efficacy [34–36].

The infective *D. maidis* adults requires a feeding period of around 1 h to transmit the spiroplasma [12]. Therefore, understanding the physicochemical properties of insecticides is crucial, as active ingredients with rapid action that disrupt or paralyze insect feeding can help reduce the risk of pathogen transmission to the plant. The insecticides evaluated in this study were selected due to their systemic action and for targeting the insect's nervous system during feeding. Those in the neonicotinoid and pyrethroid chemical groups induce paralysis and death through a knockdown effect [37]. Pyrethroids act on sodium channels, causing them to remain open for a longer duration, resulting in insect death due to hyperexcitability. Also, as neurotoxic compounds, neonicotinoids are acetylcholine agonists that bind to nicotinic receptors, opening sodium channels. Nevertheless, the molecules are not immediately degraded, leading to hyperexcitation of the nervous system and subsequent insect death [37]. In contrast, carbamates paralyze nerves and muscles by inhibiting the acetylcholinesterase enzyme [38,39]. The molecular analyses of the plants at time 0 allowed us to verify whether pathogen transmission occurred, relate the mortality rate to disease transmission, and validate the assessment scales used represented in Figure 5. Imidacloprid/thiodicarb was the insecticide that provided 100% protection by preventing spiroplasma transmission. The insecticides thiamethoxam and thiophanate-methyl + fipronil/thiamethoxam reduced the transmission rate of spiroplasma to 20% and 10% of the plants, respectively (Figure 6).

We hypothesize that these products, in addition to showing high initial mortality in leafhoppers, reduce the feeding time of the insects, contributing to the decrease in spiroplasma transmission. However, the products were only effective in preventing symptoms up to the V2 stage. This result is consistent with expectations for neonicotinoids, carbamate, and pyrethroid, which typically have a residual efficacy period from 15 to 21 days after plant emergence, depending on the soil, climatic conditions, and type of application [13,40,41]. After V2, however, with the gradual reduction of the residual effect of the products, seed insecticide treatments were not effective, and symptoms of CSS were observed in the plants, leading to yield losses.

CDSC have a lengthy developmental cycle in plants. The pathogens involved (phytoplasma and spiroplasma) require several months to multiply and reach a titer within the plant that can lead to significant economic losses. Consequently, the earlier the plants become infected, the greater the potential losses are in production [42]. In this study, we observed that the earlier the exposure of plants was to infective leafhoppers, the greater the reduction was in productivity-related parameters (Figure 6A,B), due to the longer period the pathogen had to multiply within the plant, allowing for phloem colonization and interrupting nutrient translocation [40,41]. Thus, the more effective the seed treatment was, the lower the impact was on productivity parameters [43].

The efficiency of pathogen transmission by the vector insect is not necessarily the most relevant factor from an epidemiological perspective. Pathogen transmission and disease spread depend on various factors, such as the infectivity rate of the insect population and its abundance, as well as other characteristics like biotic potential, geographic distribution, dispersal capability, and the period during which host plants are most susceptible [44]. This suggests that it is important to avoid high densities of leafhoppers in corn fields, especially during the early growth stages (VE-V4); however, this protection is discontinued at later growth phases. Delayed control of the insect vector, leading to population growth, can result in significant yield losses due to CSDC. Seed treatment, combined with good agricultural practices, is necessary to reduce the population density of leafhoppers in the field, ensuring a lower risk of the occurrence of CSDC. Finally, the higher mortality of infective leafhoppers carrying spiroplasma, compared to healthy leafhoppers, when exposed to chemical products is a novel finding that may open new research directions for future studies.

5. Conclusions

The seed treatments were effective in the initial evaluation, indicating that their protective effect persisted for approximately 5 days following the emergence of the second fully expanded leaf, or roughly 11 days after sowing. Among the evaluated products, the combination of imidacloprid and thiodicarb proved to be the most effective, providing complete (100%) protection against *Spiroplasma* infection. In contrast, thiamethoxam and the mixture of thiophanate-methyl + fipronil/thiamethoxam conferred partial protection, preventing the expression of *Spiroplasma*-associated symptoms in 80% and 90% of the plants, respectively. This has important implications for field experiments, especially for managing the insect stunting complex (CSDC). It is important to consider the possibility of reinfection and migration of healthy leafhoppers that survived seed treatment, which may require additional control measures to ensure continued protection of corn plants. Insect survival in response to control tactics may be altered, and therefore the recommendations need to be revisited.

Supplementary Materials: The following supporting information can be downloaded at: https://www.mdpi.com/article/10.3390/insects16070713/s1, Figure S1: Polymerase chain reaction (PCR) at time 0. Each number represents a treatment replicate. The samples 1–5 (=Control), 6–10 = Imidacloprid/Thiodicarb, 11–15 = Lambda-Cyhalothrin/Thiamethoxam, 16–20 = Imidacloprid, 21–25 = Clothianidin, 26–30 = Thiamethoxam, and 31–35 = Thiofanate-Methyl + Fipronil/Thiamethoxam. (M: 1 kb plus DNA ladder (Invitrogen®); C+: positive control (Spiroplasma) and C.−: negative control (all components of the PCR reaction 17 except DNA). Table S1: Rank of leafhopper daily mortality (%) after different insecticide treatment during five days in 0, 5, 10, 15 and 20 days after infestation.

Author Contributions: A.C.M.R., V.M.M., P.S.P., I.R.d.O., D.D.S.-A., L.V.C., M.A.M.F., C.M.O., D.D.R. and S.M. contributed to the study conception, data collection, and analysis. The first draft of the manuscript was written by A.C.M.R. and S.M. And all authors commented on previous versions of the manuscript. All authors have read and agreed to the published version of the manuscript.

Funding: This research was financially supported by the Embrapa/CropLife Brasil project (#10.23.00.143.00.00) and the National Council for Scientific and Technological Development (Universal—#407542/2021-3), FAPEMIG (AQP-00688-24).

Data Availability Statement: The original contributions presented in the study are included in the article/Supplementary Material. Further inquiries can be directed to the corresponding authors.

Acknowledgments: We thank the Brazilian Agricultural Research Corporation (Embrapa Maize and Sorghum) and the 'Minas Gerais State Research Support Foundation' (FAPEMIG) and 'National Council for Scientific and Technological Development' (CNPq).

Conflicts of Interest: Authors Ana Carolina M. Redoan, Poliana S. Pereira, Ivênio R. de Oliveira, Dagma D. Silva-Araújo, Luciano V. Cota, Charles M. Oliveira, Diego D. Rafael amd Simone Mendes were employed by the company Brazilian Agricultural Research Corporation (Embrapa). The remaining authors declare that the research was conducted in the absence of any commercial or financial relationships that could be construed as a potential conflict of interest.

References

1. Virla, E.G. *Dalbulus maidis* (Hemiptera: Cicadellidae), vetor del "achaparramiento o raquitismo" del maíz. Aspectos biológicos más relevantes, com especial referência a los conocimientos generados em Argentina. *Miscelanea* **2024**, *152*, 50. [CrossRef]
2. Oliveira, C.M.; Sabato, E.O.; Barros, B.A. Natural infectivity of *Dalbulus maidis* populations and incidence of corn stunt and virus diseases in maize over time. Trop. *Plant Path.* **2023**, *48*, 575–580. [CrossRef]
3. Oliveira, C.M.; Lopes, J.R.S. Técnicas para criação da cigarrinha-do-milho e inoculação de molicutes e vírus em planta. In *Doenças em Milho: Molicutes, Vírus, Vetores, Mancha por Phaeosphaeria. Brasília, DF: Embrapa Informação Tecnológica*; de Oliveira, E., de Oliveira, C.M., Eds.; Embrapa Milho e Sorgo: Sete Lagoas, Brazil, 2004; pp. 89–116.
4. Madden, L.V.; Nault, L.R.; Heady, S.E.; Styer, W.E. Effect of maize stunting mollicutes on survival and fecundity of *Dalbulus* leafhopper vectors. *Ann. App. Biol.* **1984**, *105*, 431–441. [CrossRef]
5. Jones, T.L.; Medina, R.F. Corn Stunt Disease: An Ideal Insect–Microbial–Plant Pathosystem for Comprehensive Studies of Vector-Borne Plant. *Dis. Corn. Plants.* **2020**, *9*, 747–763. [CrossRef]
6. Oliveira, C.M.; Frizzas, M.R. Eight decades of *Dalbulus maidis* (DeLong & Wolcott) (Hemiptera, Cicadellidae) in Brazil: What we know and what we need to know. *Neot. Entomol.* **2021**, *51*, 1–17. [CrossRef]
7. Oliveira, C.M.; Sabato, E.O. (Eds.) *Diseases in Maize: Insect Vectors, Mollicutes and Viruses*; Embrapa: Brasília, Brazil, 2017; 278p.

8. Ebbert, M.A.; Lowell, R. Nault. Survival in *Dalbulus* leafhopper vectors improves after exposure to maize stunting pathogens. *Entomol. Exp. App.* **2001**, *100*, 311–324. [CrossRef]
9. Ebbert, M.A.; Nault, L.R. Improved Overwintering Ability in *Dalbulus maidis* (Homoptera: Cicadellidae) Vectors Infected with (Mycoplasmatales: Spiroplasmataceae). *Environ. Entomol.* **1994**, *23*, 634–644. [CrossRef]
10. Christensen, M.N.; Axelsen, K.B.; Nicholaisen, M.; Schulz, A. Phytoplasmas and their interactions with hosts. *Trends in Plant Sci.* **2005**, *10*, 526–535. [CrossRef]
11. Gonzales, V.; Gámez, R. Algunos factores que afectam la transmisión del virus del rayado fino del maíz por *Dalbulus maidis* (DeLong & Wolcott). *Turrialba* **1974**, *24*, 51–57.
12. Alivizatos, A.S.; Markham, P.G. Multiplication of corn stunt spiroplasma in *Dalbulus maidis* and transmission in vitro, following injection. *Ann. Appl. Biol.* **1986**, *108*, 545–554. Available online: https://onlinelibrary.wiley.com/doi/pdf/10.1111/j.1744-7348.1986.tb01993.x (accessed on 26 May 2024).
13. Oliveira, C.M.D.; Oliveira, E.D.; Canuto, M.; Cruz, I. Eficiência de inseticidas em tratamento de sementes de milho no controle da cigarrinha *Dalbulus maidis* (Hemiptera: Cicadellidae) em viveiro telado. *Cienc. Rural* **2008**, *38*, 231–235. [CrossRef]
14. *Agrofit: Sistema de Agrotóxicos Fitossanitários*; MAPA: Brasília, DF, Brazil, 2024. Available online: http://agrofit.agricultura.gov.br/agrofit_cons/principal_agrofit_cons (accessed on 12 May 2024).
15. Meneses, A.R.; Querino, R.B.; Oliveira, C.M.; Maia, A.H.N.; Silva, P.R.R. Seasonal and vertical distribution of *Dalbulus maidis* (Hemiptera: Cicadellidae) in Brazilian corn fields. *Fla. Entomol.* **2016**, *99*, 750–754. [CrossRef]
16. Virla, E.G.; Coll Araoz, M.V.; Luft-Albarracin, E. Estimation of direct damage to maize seedlings by the corn leafhopper, Dalbulus maidis (Hemiptera: Cicadellidae), under different watering regimes. *Bull. Entomol. Res.* **2021**, *111*, 438–444. [CrossRef] [PubMed]
17. Ritchie, S.; Hanway, J.J. *How a Corn Plantdevelops*; Special Report; Iowa State University of Science and Technology, Cooperative Extension Service: Ames, IA, USA, 1989; Volume 48, 24p. Available online: https://publications.iowa.gov/id/eprint/18027 (accessed on 25 June 2025).
18. Silva, R.G.; Galvão, J.C.C.; Miranda, G.V.; Oliveira, E. Controle genético da resistência aos enfezamentos do milho. *Pesqui. Agropecu. Bras.* **2003**, *38*, 921–928. [CrossRef]
19. Saghai-Maroof, M.A.; Soliman, K.M.; Jorgensen, R.A.; Allard, R. Ribosomal DNA spacer-length polymorphisms in barley: Mendelian inheritance, chromosomal location, and population dynamics. *Proc. Natl. Acad. Sci. USA* **1984**, *81*, 8014–8018. [CrossRef]
20. Sousa, S.M.; Barros, B.A. *Detecção molecular de molicutes em milho In Doenças em Milho: Insetos-Vetores Molicutes e Virus*; Oliveira, C.M., Sabato, E.O., Eds.; Embrapa: Brasília, Brazil, 2017; pp. 25–34.
21. Barros, T.S.L.; Davis, R.E.; Resende, R.O.; Dally, E.L. Design of a polymerase chain reaction for specific detection of corn stunt spiroplasma. *Plant Dis.* **2001**, *85*, 475–480. [CrossRef]
22. R Development Core Team R. *A Language and Environment for Statistical Computing*; R Foundation for Statistical Computing: Vienna, Austria, 2024.
23. Gandrud, C. *Reproducible Research with R and RStudio*, 3rd ed.; CRC Press: Boca Raton, FL, USA, 2020; 276p.
24. Machado, E.P.; Souza, E.V.; Dias, G.S.; Sacilotto, M.G.; Omoto, C. Is insecticide resistance a factor contributing to the increasing problems with *Dalbulus maidis* (Hemiptera: Cicadellidae) in Brazil? *Pest Manag. Sci.* **2024**, *80*, 5120–5130. [CrossRef]
25. Alves, A.P.; Parody, B.; Barbosa, C.M.; de Oliveira, C.M.; Sachs, C.; Sabato, E.d.O.; Gava, F.; Daniel, H.; de Oliveira, I.R.; Foresti, J.; et al. Guia de Boas Práticas para o Manejo dos Enfezamentos e da Cigarrinha-do-Milho. Folhetos Embrapa Cerrados. 2020. 34p. Available online: https://www.embrapa.br/busca-de-publicacoes/-/publicacao/1129511/guia-de-boas-praticas-para-o-manejo-dos-enfezamentos-e-da-cigarrinha-do-milho (accessed on 30 July 2024).
26. Cota, L.V.; Oliveira, I.R.; Silva, D.D.; Mendes, S.M.; Costa, R.V.; Souza, I.R.P.; Silva, A.F. Manejo da Cigarrinha e Enfezamentos na Cultura do Milho. Cartilha Embrapa Milho e Sorgo. 2021. 16p. Available online: https://www.infoteca.cnptia.embrapa.br/infoteca/bitstream/doc/1130346/1/Cartilha-Manejo-cigarrinha-enfezamentos-milho.pdf (accessed on 25 July 2024).
27. Oliveira, C.M.; Oliveira, E.; Canuto, M.; Cruz, I. Chemical control of corn leafhopper and incidence of corn stunting diseases caused by mollicutes. *Pesqui. Agropecu. Bras.* **2007**, *42*, 297–303. (In Portuguese) [CrossRef]
28. Neves, T.N.C.; Josemar, F.; Silva, P.R.; Alves, E.; Rocha, R.; Oliveira, C.; Picanço, M.C.; Pereira, E.J.G. Insecticide seed treatment against corn leafhopper: Helping protect grain yield in critical plant growth stages. *Pest Manag. Sci.* **2021**, *78*, 1482–1491. [CrossRef]
29. Xavier, J.M.; Tonello, A.P. Treatments to control the corn scarlet (*Daubulus maidis*): A literature review. *J. Exact Sci.* **2023**, *39*, 36–41.
30. Martins, G.; Toscano, C.; Tomquelski, G.; Maruyama, W.I. Eficiência de inseticidas no controle de *Dalbulus maidis* (Hemiptera: Cicadellidae) na cultura do milho. *Rev. Caatinga* **2008**, *21*, 196–200.
31. Ruegger, D.G. Efeito de Inseticidas Sobre Duas Populações da Cigarrinha-do-Milho, *Dalbulus maidis* (DeLong & Wolcott) (Hemiptera: Cicadellidae). (2019). Trabalho de Conclusão de Curso—Escola Superior de Agricultura Luiz de Queiroz,

Universidade de São Paulo, Piracicaba, 2019. Available online: https://bdta.abcd.usp.br/item/003171602 (accessed on 23 July 2024).
32. Ozbek, E.; Miller, S.A.; Meulia, T.; Hogenhout, S.A. Infection and replication of *Spiroplasma Kunkelii* (Class: Mollicutes) in midgut and Malpighian tubules of the leafhopper *Dalbulus maidis*. *J. Invert. Path.* **2003**, *82*, 167–175. [CrossRef]
33. Kwon, M.O.; Wayadande, A.C.; Fletcher, J. *Spiroplasma citri* movement into the intestines and salivary glands of its leafhopper vector, *Circulifer tenellus*. *Phytopathology* **2007**, *89*, 1144–1151. [CrossRef]
34. Jeschke, P.; Nauen, R.; Schindler, M.; Elbert, A. Overview of the Status and Global Strategy for Neonicotinoid. *J. Agric. Food Chem.* **2011**, *59*, 2897–2908. [CrossRef]
35. Bonmatin, J.M.; Giorio, C.; Girolami, V.; Goulson, D.; Kreutzweiser, D.P.; Krupke, C.; Liess, M.; Long, E.; Marzaro, M.; Mitchell, E.A.; et al. Environmental fate and exposure; neonicotinoids and fipronil. *Environ. Sci. Pollut. Res. Int.* **2015**, *22*, 35–67. [CrossRef]
36. Alford, A.; Krupke, C.H. Correction: Translocation of the neonicotinoid seed treatment clothianidin in maize. *PLoS ONE* **2017**, *12*, e0186527. [CrossRef]
37. Omoto, C. Modo de ação de inseticidas e resistência de insetos a inseticida. In *Bases e Técnicas no Manejo de Insetos*; Guedes, J.C., Costa, I.D., Castiglioni, E., Eds.; UFSM/CCR/DFS: Santa Maria, CA, USA, 2000; pp. 31–49.
38. Bhirud, K.M.; Pitre, H.N. Bioactivity of systemic insecticides in corn: Relationship to leafhopper vector control and corn stunt disease incidence. *J. Econ. Entomol.* **1972**, *65*, 1134. [CrossRef]
39. Douglas, M.R.; Tooker, J.F. Large-scale deployment of seed treatments has driven rapid increase in use of neonicotinoid insecticides and preemptive pest management in 15 U.S. field crops. *Environ. Sci. Technol.* **2015**, *49*, 5088–5097. [CrossRef]
40. Hruska, A.J.; Peralta, M.G. Maize response to corn leafhopper (Homoptera: Cicadellidae) infestation and achaparramiento disease. *J. Econ. Entomol.* **1997**, *90*, 604–610.
41. Virla, E.G.; Díaz, C.; Carpane, P.; Laguna, I. Estimación preliminar de la disminución en la producción de maíz causada por el Corn Stunt Spiroplasma (CSS) en Tucumán, Argentina. *Bol. Sanid. Veg. Plag.* **2004**, *30*, 257–267.
42. Sabato, E.O. Corn stunting diseases. In *Diseases in Maize: Insect Vetors, Mollicutes and Viruses*; Oliveira, C.M., Sabato, E.O., Eds.; Embrapa Informação Tecnológica: Brasília, Brazil; Embrapa Milho e Sorgo: SeteLagoas, Brazil, 2017; pp. 11–23.
43. Lopes, B.R.; Faria, M.; Oliveira, C.M. Susceptibility of *Dalbulus maidis* to-pathogenic fungi: Unveiling the protective role of brochosomes and self-cleaning behavior. *J. Pest Sci.* **2025**, *98*, 759–768. [CrossRef]
44. Jeger, M.J. The epidemiology of plant virus disease: Towards a new synthesis. *Plants* **2020**, *9*, 1768. [CrossRef]

insects

MDPI

Article

Performance of the Fall Armyworm, *Spodoptera frugiperda* (Lepidoptera: Noctuidae), over Three Generations on Four Maize Cultivars

Bo Zhang [1], Jing Yi [1], Yan Yan [1], Yirui Wang [1], Yana Xue [1], Haiwang Yan [1], Meifeng Ren [1], Daqi Li [1], Guoping Li [2] and Junjiao Lu [1,*]

[1] Shanxi Key Laboratory of Bioagent Utilization and Eco-Pesticide Innovation, College of Plant Protection, Shanxi Agricultural University, Taiyuan 030031, China; 15133420913@163.com (B.Z.); yijingi2023@163.com (J.Y.); 18135992970@163.com (Y.Y.); 17835878261@163.com (Y.W.); 13810605632@163.com (Y.X.); yhw15245958939@163.com (H.Y.); sxzbsrmf@163.com (M.R.); daqi_li@sxau.edu.cn (D.L.)

[2] Institute of Plant Protection, Henan Academy of Agricultural Sciences, Zhengzhou 450002, China; liguoping1976@163.com

* Correspondence: lujunjiao@sxau.edu.cn; Tel.: +86-351-7117238

Simple Summary

The fall armyworm is a major pest that causes severe damage to maize crops in China, threatening food security. This study examined how four cultivars of maize—sweet, waxy, common, and silage—affected pest feeding, egg laying, and population growth over three generations. We found that newly hatched fall armyworms initially preferred sweet maize, but over time, they increasingly favored the cultivar of maize on which they were born. Although the pests could survive and reproduce on all four maize cultivars, their growth and reproduction rates varied significantly. Sweet maize supported the fastest population growth, highest survival rates, and largest egg numbers, whereas silage maize led to the lowest populations, manifested by fewer eggs and smaller pupae. By the third generation, pests developed faster across all maize types. These findings will help farmers choose maize varieties that are less likely to support pest outbreaks—such as silage maize—and avoid high-risk options, such as sweet maize. This knowledge aids in reducing crop losses, protecting maize yields, and safeguarding food production for communities.

Abstract

The fall armyworm (FAW), *Spodoptera frugiperda* (J.E. Smith), is a highly destructive pest that poses serious threats and causes significant losses to the production of maize in China. This study evaluated the feeding and oviposition preferences of *S. frugiperda* when reared on four maize cultivars—sweet, waxy, common, and silage—across three consecutive generations. It also compared population adaptability among these cultivars and analyzed population parameters between the F1 and F3 generations. The findings revealed that all four F1 generation populations showed a preference for feeding and oviposition on sweet maize. However, over time, *S. frugiperda* exhibited a stronger preference, in terms of feeding and oviposition behaviors, for the natal host plant across three consecutive generations of rearing. The fall armyworm completed its life cycle and oviposited on all four maize varieties over three generations. The sweet cultivar population had the highest intrinsic rate of increase, finite rate of increase, net reproductive rate, larval survival rate, pupation rate, eclosion rate, fecundity, and pupal weight, while the silage cultivar population had the shortest larval stage, pre-adult stage, and adult lifespan and the pupal weight and the fecundity were the lowest. Overall, the population fitness was the highest on the sweet cultivar, and the lowest on the silage cultivar. Compared with F1, the F3 generation of the FAW had a significantly shorter developmental duration in four maize cultivars. Except for the waxy maize cultivars, the fecundity of the other three cultivars did not differ significantly between F1 and F3. This study provides fundamental information on the trend of fall armyworm population changes in maize fields and serves as a reference for rational maize cultivar planting decisions.

Keywords: insect–plant interactions; pest control; agricultural entomology; crop protection; control technology

1. Introduction

The fall armyworm, *Spodoptera frugiperda* (J.E. Smith) (Lepidoptera: Noctuidae), is native to tropical and subtropical regions in the Americas. In December 2018, the FAW was first detected in Yunnan Province, China [1], and quickly spread to 26 provinces, autonomous regions, and municipalities, posing a serious economic threat to China's agricultural production [2]. The FAW is an obligatory migratory insect with a strong long-distance migration ability, capable of moving hundreds of kilometers between different regions and host plants. It can exist continuously throughout the year in the tropical and subtropical regions of southern China [3]; however, when the temperature rises in spring, it migrates seasonally to the north with the East Asian and Indian monsoons [4]. Due to its lack of diapause capacity and sensitivity to cold, it cannot survive at extremely low temperatures and returns south after autumn [5].

The FAW is a herbivorous pest with a broad host range, including 353 host plant species across 76 plant families, predominantly Gramineae, followed by Compositae and Leguminae [6]. The FAW evolved into two distinct haplotypes in the Americas, with one type primarily utilizing maize (*Zea mays* L.), sorghum [*Sorghum bicolor* (L.)], and cotton (*Gossypium hirsutum* L.), while the other type utilizes rice (*Oryza sativa* L.) and various forage crops [7,8]. Based on *COI* and *Tpi* gene analysis, the invasive Chinese population was identified as the maize type [9]. At the end of 2018, there were 3609 authorized maize cultivars in China, with a planting area exceeding 41 million hectares in 2019 [10]. A total of 1.125 million hectares of crops in China were devastated by the FAW in 2019, resulting in economic losses for the country's maize production ranging from USD 5.4 to 47 billion annually [11,12]. The FAW prefers maize as its host plant, causing damage at any stage, particularly the young leaves and growth points. Adults prefer maize for oviposition, and their larvae have significantly higher hatching and survival rates than on other plants [2].

Host selection by phytophagous insects is a complex process that involves multiple factors. However, many insects tend to prefer natal plant species (on which they have developed) when selecting plants for feeding or oviposition [13]. Hopkins' host selection principle (HHSP) states that environmental cues experienced by insects during their early stages influence their behavioral choices in later developmental phases [14]. For example, the feeding preferences of *Perina nuda* larvae are influenced by prior feeding experience, particularly when encountering different host choices [15]. Insects' prior experiences with host plants can modify their subsequent feeding and oviposition behaviors through learning [16,17]. However, these studies primarily focused on short-term trials (specific developmental stages or a single generation) to explore the effects of host experience on insect preferences. Whether learned behaviors emerge after multiple generations of continuous feeding on the same host, or whether such prolonged feeding experiences induce preference-driven shifts in host selection behavior, has often been overlooked.

Host plant species are an important factor affecting insect population growth and outbreaks. The relative adaptability of FAW populations to different host plants may result in different local population dynamics, and the longevity, fecundity, and survival rates may vary with the host plant on which the larvae feed. The effects of various host plants on FAW fitness and population parameters are detailed in Acharya et al., Li et al., and Lu et al. [18–20]. According to various studies, there is a growing distinction in growth and development patterns among successive generations as the number of generations increases [21,22]. However, many experiments only show comparisons between the first generation on different host plants or between different generations feeding on the same variety. Nevertheless, the influence of different maize cultivars on the host plant preference (larval feeding and adult oviposition) and developmental parameters (such as stage-specific lifespan, pupal weight, survival rate, and fecundity) of multiple generations of FAW populations has not been adequately addressed. Therefore, it is necessary to conduct studies on multiple generations of FAW populations to reflect the fitness of the FAW on different cultivars and the different traits among the cultivars. This will result in more precise long-term population predictions and will provide critical data for long-term FAW management.

The age-stage, two-sex life table provides a comprehensive and accurate description of the performance of insect populations under specific experimental conditions [23]. Under laboratory conditions, the age-stage, two-sex life table method was used to analyze the FAW population that feeds on various maize cultivars and has been reared for multiple generations. The F3 generation population's growth and development indexes (including developmental duration, pupation rate, eclosion rate, and fecundity in the adult stage) on various maize cultivars will be assessed and compared to determine their relative fitness. These research findings offer a basis for making informed decisions on planting various maize cultivars and supply fundamental knowledge for comprehensive FAW management.

2. Materials and Methods

2.1. Insect

The experimental insects were raised in an artificial climate box at a temperature of 25 ± 1 °C, relative humidity of 65 ± 5%, photoperiod of 16 L: 8 D, and light intensity of 12,000 lux (No. PRX-450C; Ningbo Jiangnan Instrument Factory, Ningbo, China).

Newly hatched larvae were introduced into a plastic insect-rearing box with a calligraphy brush and fed artificial diet (Table A1 for formula component and quantity) [24]. The first- to third-instar larvae were fed in groups, while the fourth- to sixth-instar larvae were individually fed in round plastic boxes (25 mL, bottom diameter 3 cm, top diameter 4 cm, height 3.5 cm). After the larvae pupated, the pupae were collected in a round plastic box covered with sand until emergence, and newly emerged adults were placed in an insect cage (35 cm × 35 cm × 35 cm) for mating. Adults were provisioned with a 15% honey solution, and eggs were collected and placed in a ziplock bag for incubation and hatching.

Insect-rearing boxes, each containing different cultivar maize leaves (sweet, waxy, ordinary, or silage), were provided to newly hatched FAW larvae that had been fed an artificial diet. The larvae of the F1 and F2 generations were reared in groups in the insect box before the third instar and individually reared in a plastic box after the third instar. The FAWs of generation F3 were reared individually during the larval and pupal stages, and a bisexual life table was created. They were observed every 24 h, the insect boxes were cleaned, and leaves were replaced.

2.2. Host Plants

The tested maize seeds were sweet cultivar (cv. Jinchaotian), waxy cultivar (cv. Jinxiannuo 6), silage cultivar (cv. Tongli 8), and common cultivar (cv. Dafeng 30). Four different cultivars of maize seeds were planted in plastic pots (50 cm × 35 cm × 25 cm), and substrate nutrient soil was used for planting. The sowing depth was 5 cm, and the planting density was 50 plants per pot. The maize plants were cultivated under natural outdoor lighting conditions. Irrigation was performed at 2- to 3-day intervals. After the maize had grown to the 5-leaf stage, young leaves were cut for feeding and testing. None of the host plants for the test were exposed to any pesticides.

2.3. Determination of *Spodoptera frugiperda* Oviposition Preference

An oviposition preference assay was conducted using a cage experiment to evaluate the oviposition preferences of female fall armyworm adults on four maize varieties. Maize plants of the four varieties at the three-leaf stage with similar plant heights and leaf sizes were randomly arranged at the four corners of a mesh cage (40 cm × 40 cm × 40 cm, 200-mesh). A plastic Petri dish filled with cotton moistened with a 15% honey solution was placed at the center of the cage. Eight newly emerged adult pairs (1:1 sex ratio) were introduced into the cage and allowed to mate freely. The number of eggs deposited on each variety was recorded 48 h post-introduction, with data collected continuously for 5 days across three experimental replicates. Plants were watered daily, and the honey solution was replenished as needed. The formula for calculating the egg attachment rate was as follows: (number of eggs on a single maize variety/total number of eggs across the four maize varieties) × 100%. The entire experiment was conducted in a climate-controlled greenhouse at a temperature of 25 ± 1 °C, relative humidity of 65 ± 5%, and a photoperiod of 16 L: 8 D.

2.4. Determination of *Spodoptera frugiperda* Feeding Preference

Feeding preference was assessed using the leaf disk assay. Plastic Petri dishes (15 cm diameter) lined with filter paper were divided into four equal sections, each containing maize leaf segments of equal weight that were moistened to maintain hydration. Ten second-instar larvae starved for 6 h prior to the assay were placed at the center of each dish. The experiment included ten replicates per treatment. The number of larvae feeding on each variety was recorded after 24 h to calculate the feeding selection rates. The experimental conditions were identical to those described in the Section 2.1.

2.5. Construction of Life Tables of *Spodoptera frugiperda*

Using the laboratory population life table method, we evaluated the effects of continuous multi-generation rearing on different maize cultivars on the growth, development, and reproduction of FAWs. F3 generation eggs that were continuously reared for multiple generations with different maize cultivars were used as the test insects. After hatching, the leaves of the four different maize cultivars were used to feed the larvae. Newly hatched larvae were placed in plastic boxes for individual rearing. In the life table study, a total of 100 newly hatched larvae were used for each maize cultivar, each larva as a replicate. Maize leaves were replaced every day, and larval survival, instar, pupation, eclosion, and other parameters were recorded daily. The pupal weight was measured on the second day after pupation. After emergence, adults of the same day (one female and one male) were placed in a transparent plastic cup (250 mL, with a bottom diameter of 5 cm, a top diameter of 7.5 cm, and a height of 7 cm) for 1:1 pairing. The plastic cup was sealed with double-layer degreasing gauze for female adults to lay eggs. A cotton ball soaked with 15% honey water was hung in the plastic cup for adults to replenish nutrition and was replaced daily. On the second day after mating, egg laying by the adults in the plastic cup was observed and recorded. The eggs were collected daily until the adults died, and the lifespans of male and female adults were recorded. The entire test was conducted in an artificial climate box with a temperature of (25 ± 1) °C, relative humidity of (65 ± 5) %, and photoperiod of 16 L: 8 D.

2.6. Life Table Data Analysis

The data for the experiment were collected and analyzed using the TWOSEX-MS Chart 2022 software [25] (available at http://lifetablechi.com/software/, accessed on 9 January 2023). The age-stage-specific survival rate (s_{xj}), age-specific survival rate (l_x), age-stage-specific fecundity (f_{xj}), age-specific fecundity (m_x), age-specific maternity ($l_x m_x$), net reproductive rate (R_0), intrinsic rate of increase (r), finite rate of increase (λ), mean generation time (T), age-stage-specific life expectancy (e_{xj}), age-stage-specific reproductive value (v_{xj}), and other analytical data were processed and plotted [26,27]. All definitions, equations, and references are listed in Table A2.

The mean and standard errors of each parameter were calculated by using the bootstrap method supported by the TWOSEX-MS Chart software (with 100,000 bootstraps), and then the paired bootstrap test was used to conduct significance analysis of differences in the data [28].

2.7. Data Processing and Mapping

The original data source was recorded and organized using Excel 2022, and then the original data source was analyzed using the TWOSEX-MS Chart software and plotted using SigmaPlot 15.0 (SigmaPlot Software, San Rafael, CA, USA). Normality of feeding and oviposition preferences were measured using the Kolmogorov–Smirnov test in SPSS v.27.0, then one-way analysis of variance (significance level set at 0.05) was performed, followed by Duncan test comparisons to determine significant differences. Percentage data were converted to the square root of the arcsine before one-way analysis of variance (ANOVA) to meet the requirements of ANOVA.

3. Results

3.1. The Oviposition Preference of Spodoptera frugiperda on Four Maize Cultivars for Three Consecutive Generations of Feeding Experiences

The oviposition preference of the FAW populations reared for three consecutive generations on the same maize cultivar exhibited significant changes (Table 1). The F1 generation predominantly oviposited on sweet maize in all populations, but their preference gradually shifted toward the original maize cultivar that they had previously fed on as generations progressed. For instance, in the sweet cultivar and silage cultivar populations, significant differences in egg-laying rates on the original maize cultivar were observed between the F1 and F3 generations ($F = 6.00$, *d.f.* = 2,6, $p < 0.05$; $F = 6.62$, *d.f.* = 2,6, $p < 0.05$). In the waxy cultivar population, the egg-laying rate of the F3 generation on the original maize cultivar significantly differed from that of the F1 and F2 generations ($F = 5.40$, *d.f.* = 2,6, $p < 0.05$), whereas in the common cultivar population, the egg-laying rate of the F1 generation was significantly different from that of the F2 and F3 generations ($F = 5.40$, *d.f.* = 2,6, $p < 0.05$).

Table 1. Oviposition preference (mean ± SE) of *Spodoptera frugiperda* adults on four maize cultivars for three consecutive generations.

Population	Generation	Maize Variety			
		Sweet Cultivar	Waxy Cultivar	Common Cultivar	Silage Cultivar
Sweet cultivar	F1	26.5 ± 0.4 b	24.9 ± 1.2 a	23.6 ± 1.0 a	25.0 ± 0.3 a
	F2	28.7 ± 0.9 ab	24.3 ± 0.4 a	21.5 ± 0.3 a	25.5 ± 0.5 a
	F3	31.3 ± 1.4 a	22.9 ± 1.2 a	21.5 ± 0.7 a	24.3 ± 0.8 a
Waxy cultivar	F1	28.0 ± 1.1 a	23.2 ± 0.1 b	24.0 ± 0.9 a	24.8 ± 0.6 a
	F2	27.0 ± 1.0 a	23.8 ± 0.6 b	24.0 ± 0.8 a	25.2 ± 0.5 a
	F3	26.8 ± 0.8 a	26.2 ± 1.0 a	23.9 ± 1.8 a	23.2 ± 1.0 a

Table 1. *Cont.*

Population	Generation	Maize Variety			
		Sweet Cultivar	Waxy Cultivar	Common Cultivar	Silage Cultivar
Common cultivar	F1	27.8 ± 1.4 a	22.8 ± 0.7 a	25.2 ± 0.8 b	24.3 ± 0.3 a
	F2	26.3 ± 0.9 a	22.8 ± 0.9 a	28.1 ± 0.8 a	22.8 ± 1.2 a
	F3	27.0 ± 1.7 a	21.4 ± 1.1 a	28.2 ± 0.6 a	23.5 ± 1.9 a
Silage cultivar	F1	28.1 ± 2.0 a	22.7 ± 0.6 a	23.7 ± 1.0 a	25.5 ± 1.0 b
	F2	25.7 ± 0.3 a	22.2 ± 0.1 a	24.6 ± 0.7 a	27.5 ± 0.8 ab
	F3	24.1 ± 0.5 a	21.7 ± 1.4 a	23.7 ± 0.7 a	30.5 ± 1.1 a

Different letters indicate significant differences in egg attachment rates among the F1, F2, and F3 generations of the same *Spodoptera frugiperda* population on the same maize cultivar (one-way ANOVA, Duncan test, $p < 0.05$).

3.2. The Feeding Preferences of Spodoptera frugiperda Larvae on Four Maize Cultivars for Three Consecutive Generations of Feeding Experiences

The feeding preferences of FAW larvae populations continuously reared on the same maize cultivar for three generations exhibited different variations, either among different cultivars within the same generation or among different generations on the same cultivar (Figure 1). In all populations, the F1 generation larvae consistently demonstrated a feeding preference for the sweet maize cultivar. However, their preference for natal host plants gradually increases with successive generations. For example, populations reared on sweet, common, and silage cultivars showed significant differences in larval feeding preference for their natal maize cultivar in all generations ($p < 0.05$). Although no significant differences were observed among the three generations in the waxy cultivar population, the preference rate for natal hosts also exhibited an increasing trend with generational progression.

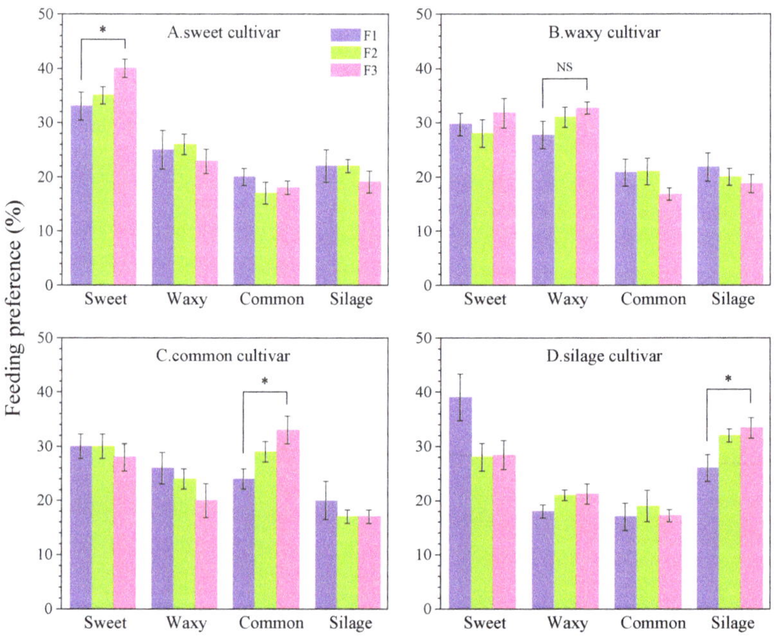

Figure 1. Feeding preferences of *Spodoptera frugiperda* larvae on 4 maize cultivars for three consecutive generations after 24 h. The data presented are mean ± SE. Statistical significance was assessed using the Duncan multiple comparison test. The error bar in the figure is SE. (* $p < 0.05$, NS $p > 0.05$).

3.3. *Effects of Different Feeding Experiences on the Developmental Duration of Spodoptera frugiperda*

All FAWs were able to complete their life cycle when reared for multiple generations with the four maize cultivars (Table 2). In the common cultivar treatment, the larval stage (18.4 d), pupal stage (9.5 d), pre-adult stage (29.2 d), and total duration (37.4 d) were longer than those in the other three treatments. This indicates that FAWs feeding on the common cultivar developed relatively slowly.

Table 2. Development time, adult longevity, and total longevity of *Spodoptera frugiperda* reared on four maize cultivars.

Developmental Stage	Maize Cultivar							
	n	Sweet Cultivar	n	Waxy Cultivar	*n*	Common Cultivar	n	Silage Cultivar
Egg (d)	99	3.0 ± 0.0 a	99	3.0 ± 0.0 a	100	3.0 ± 0.0 a	100	3.0 ± 0.0a
1st instar (d)	96	2.9 ± 0.1 a	97	2.8 ± 0.0 a	98	2.9 ± 0.1 a	94	2.6 ± 0.1 b
2nd instar (d)	92	2.1 ± 0.0 b	95	2.0 ± 0.0 b	93	2.3 ± 0.0 a	93	2.2 ± 0.1 a
3rd instar (d)	91	2.2 ± 0.9 a	90	2.2 ± 0.1 a	91	2.3 ± 0.1 a	90	2.0 ± 0.1 b
4th instar (d)	90	2.5 ± 0.1 b	89	2.5 ± 0.0 ab	91	2.7 ± 0.1 a	86	2.1 ± 0.1 c
5th instar (d)	88	2.5 ± 0.0 a	89	2.2 ± 0.0 c	91	2.4 ± 0.1 ab	86	2.3 ± 0.0 bc
6th instar (d)	87	3.1 ± 0.1 a	84	3.0 ± 0.1 a	84	2.8 ± 0.1 b	83	3.0 ± 0.1 a
Total larval duration (d)	87	15.2 ± 0.1 a	84	15.0 ± 0.1 b	84	15.4 ± 0.1 a	83	14.3 ± 0.1 c
Prepupa (d)	84	1.2 ± 0.1 c	76	1.5 ± 0.1 a	80	1.4 ± 0.1 b	78	1.4 ± 0.1 b
Pupa (d)	72	8.7 ± 0.2 b	65	8.8 ± 0.1 b	68	9.5 ± 0.1 a	64	9.0 ± 0.1 a
Total immature duration (d)	72	28.1 ± 0.2 b	65	28.3 ± 0.2 b	68	29.2 ± 0.2 a	64	27.7 ± 0.2 c
Immature survival rate (s_a) (%)	100	72 ± 4 a	100	65 ± 5 a	100	68 ± 5 a	100	64 ± 5 a
Female adult longevity (d)	39	8.5 ± 0.3 a	34	7.8 ± 0.4 a	33	8.2 ± 0.4 a	34	7.8 ± 0.3 a
Male adult longevity (d)	33	8.4 ± 0.4 a	31	7.6 ± 0.4 a	35	8.1 ± 0.3 a	30	7.2 ± 0.3 a
Female total longevity (d)	39	36.6 ± 0.3 a	34	34.7 ± 0.5 b	33	36.6 ± 0.5 a	34	34.3 ± 0.3 b
Male total longevity (d)	33	37.2 ± 0.4 a	31	37.4 ± 0.6 ab	35	38.1 ± 0.3 a	30	36.1 ± 0.4 b
Mean longevity (all individuals) (d)	100	31.4 ± 1.0 a	100	29.8 ± 1.0 ab	100	31.4 ± 1.0 a	100	28.6 ± 1.0 b

The data in the table are presented as mean ± SE. Different letters in the same row indicate significant differences among different host plants (one-way ANOVA, paired bootstrap test, $p < 0.05$). Tables 2–4 are the F3 data of FA.

3.4. *Effects of Different Feeding Experiences on Pupal Weight of Spodoptera frugiperda*

There were significant differences ($p < 0.05$) in the pupal weights of different populations when the FAW was reared with the four maize cultivars after three consecutive generations (Table 3). The pupal weight of both female and male insects was the highest on the sweet cultivar treatment (152.6 mg and 159.8 mg, respectively), and the lowest on the silage cultivar treatment (132.7 mg and 141.8 mg, respectively). The pupal weight of male moths in all treatments was higher than that of the females. With data pooled together, the pupal weight of the treatment on the sweet cultivar was still the heaviest at 155.9 mg, while the pupal weight of the treatment on the silage cultivar was the lightest at 137.3 mg.

Table 3. Pupal weight of *Spodoptera frugiperda* reared on the four maize cultivars.

Sex	Pupal Weight (mg)						
	Sweet Cultivar	Waxy Cultivar	Common Cultivar	Silage Cultivar	*F*	*d.f.*	*p*
Female	152.6 ± 2.2 a	143.1 ± 3.8 b	142.7 ± 3.8 b	132.7 ± 3.2 c	6.83	3,118	<0.05
Male	159.8 ± 3.8 a	152.6 ± 4.9 ab	155.8 ± 4.0 a	141.8 ± 3.9 b	3.50	3,109	<0.05
Total	155.9 ± 2.1 a	147.63 ± 3.1 b	149.2 ± 2.9 ab	137.3 ± 2.6 c	8.67	3,231	<0.05

The data in the table are presented as mean ± SE. (One-way ANOVA, Duncan test, $p < 0.05$.) Different letters indicate significant differences between the four cultivars of the same parameter.

3.5. Effects of Different Feeding Experiences on the Survival of *Spodoptera frugiperda*

The larval survival, pupation, and eclosion rates of the F3 generation of the FAW had different effects among the four treatments (Table 4). Compared with the other three treatments, the sweet cultivar had the highest cumulative survival, pupation, and emergence rates of larvae, which were 88.0%, 96.6%, and 85.7%, respectively. The silage cultivar treatment had the lowest cumulative survival and emergence rates of larvae, which were 83.0% and 82.1%, respectively. The waxy cultivar had the lowest pupation rate (90.5%).

Table 4. Survival rate of *Spodoptera frugiperda* reared on four maize cultivars.

Maize Cultivar	Survival Situation		
	Total Larval Survival Rate (%)	Pupation Rate (%)	Emergence Rate (%)
Sweet cultivar	88.0 ± 3.6 a	96.6 ± 2.0 a	85.7 ± 3.8 a
Waxy cultivar	84.0 ± 3.6 b	90.5 ± 3.2 d	85.5 ± 4.1 a
Common cultivar	84.0 ± 3.2 b	95.2 ± 2.3 b	85.0 ± 4.0 a
Silage cultivar	83.0 ± 3.5 c	94.0 ± 2.6 c	82.1 ± 4.4 b
F	41.97	98.37	22.06
d.f.	3,396	3,396	3,396
p	<0.05	<0.05	<0.05

The data in the table are presented as mean ± SE. (One-way ANOVA, paired bootstrap test, $p < 0.05$.) Different letters indicate significant differences between the four cultivars of the same parameter.

3.6. Effects of Different Feeding Experiences on the Reproduction of Adult *Spodoptera frugiperda*

The adult preoviposition period (APOP) of the FAW reared on the silage cultivar treatment was the longest at 4.4 d (Table 5). Compared with other treatments, the oviposition days (O_d) of the sweet cultivar treatment was the longest (O_d = 3.7), the mean fecundity of all female adults (F) and the mean fecundity of only reproductive female adults (F_r) was the highest (F = 458 eggs/female, F_r = 596 eggs/reproductive female), while the O_d of the waxy cultivar treatment was the shortest (O_d = 3.3 d), and the F of the silage cultivar treatment was the lowest (F = 371 eggs/female).

3.7. Effects of Different Feeding Experiences on Life Table Population Parameters of *Spodoptera frugiperda*

The mean generation time (T) showed a significant difference ($p < 0.05$), whereas the net reproductive rate (R_0), intrinsic rate of increase (r), and finite rate of increase (λ) were not significantly different ($p > 0.05$) among the FAW populations reared on the four maize cultivars (Table 5).

3.8. Parameters of the F1 and F3 Generations of *Spodoptera frugiperda* Feeding on Four Maize Cultivars

The larval stage, pupal stage, total preoviposition period (TPOP), and mean generation time (T) of the F3 generation of the FAW were significantly shorter than those of the F1 generation ($p < 0.05$, Table 5). The survival rate of the immature stage (S_a) and the net reproductive rate (R_0) of the F3 generation were similar to those of the F1 generation. The intrinsic rate of increase (r) and finite rate of increase (λ) in the sweet, waxy, common, and silage cultivars increased in the F3 generation, but only the increase in the silage cultivars was significant ($p < 0.05$). The fecundity per female (F_r) and mean fecundity (F) of the ovipositing female insects on the waxy cultivar decreased as the number of generations increased, while those reared on the silage cultivar increased.

Table 5. Population parameters of *Spodoptera frugiperda* F1 and F3 generations reared on the four maize cultivars.

Parameters	F1 Generation				F3 Generation			
	Sweet Cultivar	Waxy Cultivar	Common Cultivar	Silage Cultivar	Sweet Cultivar	Waxy Cultivar	Common Cultivar	Silage Cultivar
Total larval duration (d)	17.4 ± 0.1 a	17.7 ± 0.2 a	18.5 ± 0.1 a	17.6 ± 0.2 a	16.8 ± 0.1 b	16.2 ± 0.1 b	16.8 ± 0.1 b	15.4 ± 0.1 b
Pupa (d)	10.5 ± 0.1 a	10.4 ± 0.1 a	10.3 ± 0.1 a	10.4 ± 0.1 a	8.7 ± 0.1 b	8.8 ± 0.2 b	9.5 ± 0.1 b	9.0 ± 0.2 b
Immature survival rate (s_a) (%)	74 ± 4 a	70 ± 5 a	70 ± 5 a	67 ± 5 a	72 ± 4 a	65 ± 5 a	68 ± 5 a	64 ± 5 a
Mean longevity (d)	33.9 ± 1.0 a	32.8 ± 1.1 a	33.2 ± 1.2 a	32.0 ± 1.2 a	31.4 ± 1.0 a	29.8 ± 1.0 b	31.4 ± 1.0 a	28.6 ± 1.0 b
APOP (d)	4.3 ± 0.1 a	4.1 ± 0.2 a	4.2 ± 0.2 a	4.4 ± 0.2 a	3.5 ± 0.2 b	3.7 ± 0.2 a	3.4 ± 0.1 b	3.8 ± 0.1 b
TPOP (d)	34.3 ± 0.3 a	35.0 ± 0.4 a	35.4 ± 0.3 a	35.4 ± 0.4 a	31.5 ± 0.3 b	30.5 ± 0.3 b	31.9 ± 0.3 b	30.2 ± 0.3 b
Oviposition day (d)	3.7 ± 0.1 a	3.5 ± 0.2 a	3.0 ± 0.2 a	3.2 ± 0.2 a	3.7 ± 0.1 a	3.3 ± 0.2 a	3.3 ± 0.2 a	3.5 ± 0.1 a
F (all female) (eggs/female)	556 ± 28 a	531 ± 31 a	398 ± 31 a	399 ± 30 a	458 ± 46 a	382 ± 39 b	425 ± 44 a	371 ± 40 a
F_r (rep. female) (eggs/female)	603 ± 9 a	581 ± 16 a	458 ± 20 a	452 ± 19 b	596 ± 29 a	481 ± 26 b	520 ± 32 a	504 ± 17 a
R_0 (offspring)	211.14 ± 29.01 a	185.78 ± 27.56 a	151.30 ± 22.53 a	135.53 ± 21.32 a	178.70 ± 28.66 a	129.98 ± 22.39 a	140.30 ± 24.56 a	125.97 ± 22.14 a
r (d^{-1})	0.15 ± 0.04 a	0.14 ± 0.04 a	0.14 ± 0.04 a	0.13 ± 0.05 b	0.15 ± 0.05 a	0.15 ± 0.06 a	0.15 ± 0.06 a	0.15 ± 0.06 a
λ (d^{-1})	1.16 ± 0.05 a	1.15 ± 0.05 a	1.15 ± 0.05 a	1.14 ± 0.05 b	1.17 ± 0.06 a	1.16 ± 0.07 a	1.16 ± 0.06 a	1.16 ± 0.07 a
T (d)	35.8 ± 0.3 a	36.5 ± 0.4 a	36.9 ± 0.3 a	36.8 ± 0.4 a	33.8 ± 0.2 b	32.4 ± 0.3 b	33.7 ± 0.3 b	32.1 ± 0.3 b

Different letters indicate significant differences between the F1 and F3 generations of the same host plant. The data in the table are presented as mean ± SE. (One-way ANOVA, paired bootstrap test, $p < 0.05$.) (APOP: the period between adult emergence and first oviposition. TPOP: the period between egg birth and adult first oviposition.)

3.9. Age-Stage Survival Rate of *Spodoptera frugiperda* with Different Feeding Experiences

The survival rates (s_{xj}) of FAW populations continuously reared on the four maize cultivars for three generations were relatively higher than those of the first generation (Figure 2). The survival rates of the pupal stage reared on the sweet, waxy, common, and silage cultivars were 85%, 78%, 80%, and 77%, respectively. At the initial adult stage, the survival rate of female moths was higher than that of males, whereas the results in the later stages were the opposite. Therefore, the total lifespan of the FAW on the waxy cultivar and common cultivar after three generations of rearing was longer than that on the sweet cultivar and silage cultivar populations.

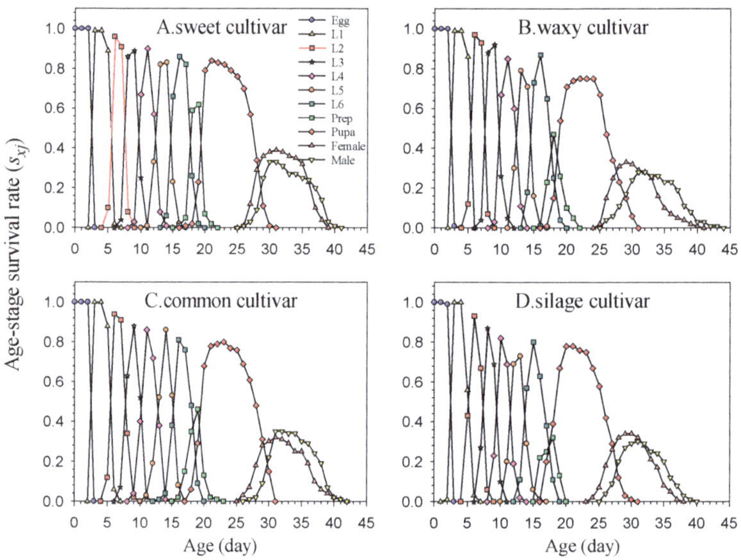

Figure 2. Age-stage survival rate (s_{xj}) of *Spodoptera frugiperda* reared on four maize cultivars. Figures 2–5 are the F3 data of FAW.

3.10. Age-Specific Survival Rate and Population Fecundity of *Spodoptera frugiperda* with Different Feeding Experiences

Age-specific fecundity (m_x) was the average number of eggs produced by the FAW population (Figure 3). During the entire development process, the female on the silage cultivar started to oviposit first, while on the common cultivar, it was the last to oviposit. The peak values of the mx curves are 85, 62, 57, and 52, respectively. The m_x value for the sweet cultivar was the highest, whereas that of the silage cultivar was the lowest. The parameter f_{x10} is the average daily egg production at age x stage 10. Here, the sweet cultivar had the highest oviposition peak, whereas the silage cultivar had the lowest oviposition peak. Finally, $l_x m_x$ is the total number of eggs laid per female at age x and can indicate the population fertility, considering the survival rate. The $l_x m_x$ peak values for FAWs feeding on the sweet cultivar, waxy cultivar, common cultivar, and silage cultivar were 52.36, 37.16, 36.89, and 28.62, respectively, and the first time of reproduction on the sweet cultivar was the fastest among them.

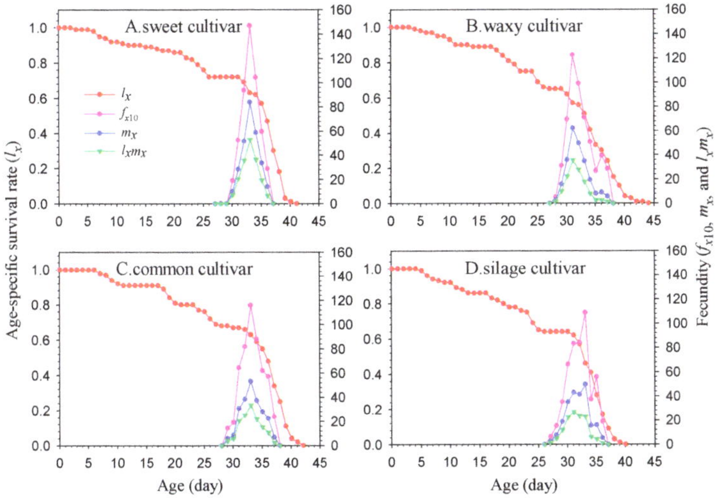

Figure 3. Age-specific survival rate (l_x), female age-specific fecundity (f_{x10}) (female adult is the 10th stage), age-specific fecundity (m_x), and age-specific net maternity (l_xm_x) of *Spodoptera frugiperda* reared on the four maize cultivars.

3.11. The Expected Lifespan of the Population of Spodoptera frugiperda with Different Feeding Experiences

Age-stage-specific life expectancy (e_{xj}) represents the expected lifespan of each individual (Figure 4). The expected value curve for the silage cultivar dropped the fastest, and the expected value of the starting lifespan at each stage was lower than that of other populations, indicating that the growth rate of the FAW on the silage cultivar was faster than that of other populations, which was the same as the result of the larval development duration on the silage cultivar. In addition, the expected lifespan values of male adults in the four FAW populations were higher than those of female adults.

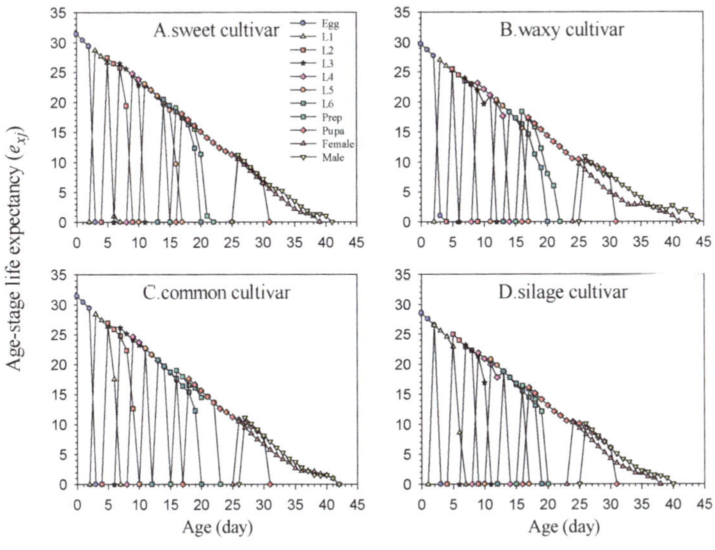

Figure 4. Population life expectancy (e_{xj}) of *Spodoptera frugiperda* reared on four maize cultivars.

3.12. The Population Reproductive Value of *Spodoptera frugiperda* with Different Feeding Experiences

The age-stage-specific reproductive value (v_{xj}) refers to the average contribution of an individual to future population growth (Figure 5). The reproductive value increased with the larval instar number. The initial egg-laying reproductive values of the FAW population reared on the sweet, waxy, common, and silage cultivars were 1.1656, 1.1623, 1.1578, and 1.1625, respectively, which were consistent with the finite rate of increase (λ) in the life table parameters. The reproductive peaks of the FAW population appeared at 32–34 days among the four treatments, which were (sweet cultivar, 32 days, 325 eggs/female), (waxy cultivar, 30 days, 280 eggs/female), (common cultivar, 31 days, 302 eggs/female), and (silage cultivar, 29 days, 246 eggs/female), respectively. The reproductive value peak of the sweet cultivar was the highest but appeared the latest, and the peak of the silage cultivar was the lowest but appeared the earliest. The egg-laying days of female adults on the sweet, waxy, common, and silage cultivars were 11, 13, 12, and 13 days, respectively. The peak reproductive value of female adults on the sweet cultivars was the highest, but the reproduction duration was the shortest.

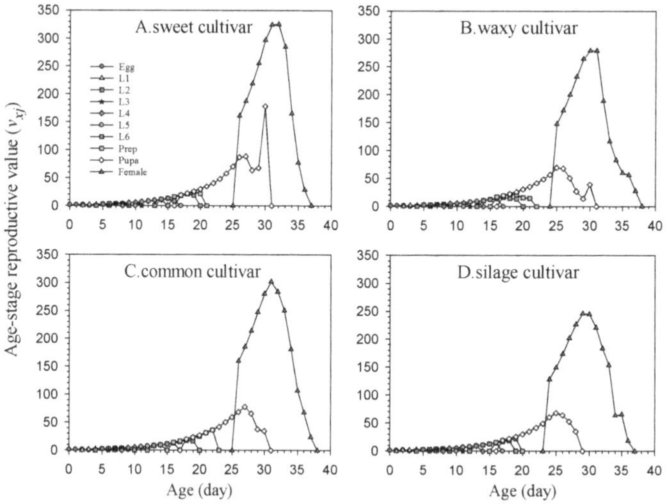

Figure 5. Population reproduction value (v_{xj}) of *Spodoptera frugiperda* reared on four maize cultivars.

4. Discussion

The results of this study revealed that F1 generation fall armyworm populations reared on the four maize cultivars exhibited larval and adult preferences for sweet maize. Relevant studies have shown that substrate characteristics, such as sweetness and hardness, influenced oviposition choices, with a positive correlation observed between oviposition preference and glucosinolate content in host plants, which may also explain the fall armyworm's preference for sweet maize [29,30]. With successive generations, the preference hierarchy of the FAW for the four maize cultivars changed significantly and consistently as larvae and adults gradually increased their feeding and oviposition activities on their original host plants. This indicated that the FAW had learning behavior, and feeding experience could induce preference-driven host selection, aligning with Hopkins' host selection principle (HHSP). Female onion flies (*Delia antiqua*) exhibit a strong preference and deposit more eggs on their original host plant [13]. Conversely, when female tobacco hornworms (*Manduca sexta* L.) completed a single oviposition event on one of two host plants (*Datura wrightii* or *Nicotiana attenuata*), they immediately selected the previously experienced plant foliage when re-exposed to both options, demonstrating that a single oviposition event can drive subsequent oviposition preference changes [31].

Host plants play an important role in the growth, development, and reproduction of insects, and suitable hosts can improve the growth rate, survival rate, and fecundity of insect offspring [32]. This experiment studied the effects of feeding on different maize cultivars on the growth, development, and reproduction of successive generations of FAW until the F3 generation. These results showed significant effects on the developmental duration of each insect stage, pupal weight, survival rate, and oviposition of the FAW population reared on the different maize cultivars. FAWs feeding on six rice cultivars resulted in significant differences in larval development duration, pupal duration, pupation rate, adult lifespan, and egg production [33]. These results showed that the larval stage and adult lifespan of FAWs reared on sweet maize were both longer than those reared on waxy maize [34]. Similarly, Zhang et al. indicated that special maize cultivars are more suitable for the growth and development of FAWs than ordinary maize cultivars [35]. Compared to common maize cultivars, these populations have stronger adaptability to waxy maize, which is consistent with the results of our study. However, the larval stage and adult lifespan of FAWs feeding on waxy maize are longer than those on common maize, which is contrary to the results of this study, and might be caused by differences in cultivars or rearing patterns [10,36]. Generally, when an insect on a certain host plant has higher fitness, it will have higher rates of development, survival, and reproduction. Although the growth and development speed are faster when the FAW feeds on wheat (*Triticum aestivum* L.) than on maize, its food utilization efficiency and population reproduction ability were all lower [37], which is similar to our results showing that the larval development duration of the silage maize population was the shortest, but the survival, pupation, and eclosion rates were the lowest.

The state of the pupa reflects the adaptability of the larva to a particular host or environment, and the weight of the pupae reflects the insect's appetite for the host plant [38]. The pupal weight and fecundity of female lepidopteran adults are positively correlated with their adaptive potential [35]. The results of our study showed that the pupal weight of the FAW differed significantly among the four maize cultivars (sweet maize > common maize > waxy maize > silage maize). Meanwhile, the larval survival rate, pupal weight, and egg production of female adults of FAWs fed sweet maize were significantly higher than those fed waxy maize [34]. Additionally, the pupal weight and egg production of FAWs fed on common maize were both higher than those of waxy maize [10]. These results are consistent with those of our study. The female pupal weight on special maize cultivars was higher than that of common maize cultivars [35]; however, in our study, the pupal weight of waxy maize was lower than that of common maize, which may be related to the nutritional quality or resistance among different maize cultivars. Other relevant studies have shown that the change in pupal weight may be related to the amino acid content among different cultivars. Cultivars with strong resistance generally have higher glutamic acid content and lower tyrosine content, and the resulting pupal weight is lighter [39].

The biological parameters R_0, r, λ, and T indicate the growth, development, reproduction, and survival changes in insects and the population growth ability in a specific environment [40]. There were significant differences in the population parameters of FAWs among different maize cultivars. Our results showed that the R_0, r, and λ values were highest on sweet maize. Similarly, the intrinsic rate of increase and net reproductive rate of FAWs feeding on sweet maize were also higher than those on waxy maize [34]. In contrast, Zhang et al. reported that the net reproductive rate and intrinsic rate of growth were waxy maize > sweet maize > common maize, which was different from the results of our study [35]. This may be caused by different generations of FAW on the same host.

FAWs feeding on plants with secondary substances that are different from hosts for multiple generations will cause changes in the activities of some enzymes in the insects and have an impact on the growth and development of larvae [41]. When the FAW feeds on *Vicia villosa* Roth, as the number of generations increases, its adaptability gradually decreases until it cannot complete a subsequent generation or maintain its population at the fourth generation [22]. When returned to the original host maize, the offspring displayed reduced performance and loss of adaptation to their host. Although Bt cotton had a negative impact on *Spodoptera exigua* Hübner for three continuous generations, the survival rate and fecundity of adults increased significantly, and the lipase and trypsin activities of the third generation were significantly lower than those of the first generation; however, the activities of carboxylesterase and acetylcholinesterase were significantly higher than those of the first generation [42]. This indicates that the enhancement of its adaptability may be closely related to the enhancement of detoxifying enzyme activities rather than digestive enzymes. Therefore, further verification is needed in the next research. Similarly, the pupal weight, survival rate, fecundity, relative growth rate, and relative digestion rate of the third generation were all significantly higher than those of the first generation when the larvae of *Helicoverpa armigera* Hübner fed on high-gossypol cultivars [21]. The developmental duration was prolonged with an increase in successive rearing generations when the guava fruit fly (*Bactrocera correcta* Bezzi) was continuously reared indoors on an artificial diet [43]. After being fed rape pollen, the F1 and F2 generations of *Micraspis discolor* showed a lower survival rate and female ratio, but the F3 and F4 generations had higher survival rates, female ratios, and weights, suggesting that *M. discolor* gradually adapted to the pollen [44].

Developmental duration is critically important for insect survival, as prolonged exposure to natural environments elevates the risk of biotic (e.g., pathogens and natural enemies) and abiotic (e.g., natural disasters and adverse environmental conditions) stressors, which may pose critical threats to their viability. Because the life table of the F1 generation of FAWs has already been studied in our previous experiments, we measured the parameters of the F3 generation in our study. By comparing our results with those of the F1 generation, it can be seen that the developmental duration of the F3 generation of FAWs was shortened significantly [45]. This indicates that the FAW can adapt to diverse plant species through evolution, shortened developmental duration, and mitigated adverse factors affecting its development, thereby enhancing its adaptability to different ecological regions.

In conclusion, since FAWs are a type of polyphagous pest that can damage different crops and cultivars within a certain area, at the same time, long-term cultivation of a single plant or cultivar may also aggravate the damage situation of FAWs. Therefore, it is necessary to implement habitat regulation technology such as intercropping instead of a single crop or variety in the field, or plant protective row plants at the edge of the field, to reduce the possibility of multiple generations of FAWs completing their development on the same plant. These strategies will reduce the loss of crop production due to the presence of FAWs.

Author Contributions: Conceptualization, J.L. and H.Y.; investigation, B.Z., J.Y., Y.Y., and H.Y.; data analysis, graph and chart creation, and manuscript writing, B.Z., Y.W., M.R., and D.L.; revision of the manuscript, B.Z., Y.X., and J.L.; funding acquisition, J.L. and G.L. All authors have read and agreed to the published version of the manuscript.

Funding: This research was funded by the National Key Research and Development Program of China, grant number 2021YFD1400705-7 and the earmarked fund for Modern Agro-industry Technology Research System of Shanxi, grant number 2024CYJSTX01-18.

Data Availability Statement: The original contributions presented in this study are included in the article. Further inquiries can be directed to the corresponding author.

Conflicts of Interest: The authors declare no competing interests.

Appendix A

Table A1. Artificial diet formulation.

Component	Quantity
Wheat germ powder	280 g
Soy protein powder	90 g
Yeast powder	35 g
Agar	25 g
Vitamin B complex	0.2 g
Cholesterol	12 g
Sorbic acid	2 g
Ascorbic acid	12 g
Methyl p-hydroxybenzoate	5 g
Formaldehyde	4 mL
Penicillin	0.2 g
Distilled water	1500 mL

Table A2. The definition, equation, and reference of population parameters used in this study.

Parameter	Equation	Definition and Reference
Age-specific survival rate (l_x)	$l_x = \sum_{j=1}^{m} s_{xj}$	The probability that a newborn offspring survives to age x. It includes female, male, and those that died in the pre-adult stages; m is the number of stages [40].
Age-specific fecundity (m_x)	$m_x = \sum_{j=1}^{m} s_{xj} f_{xj} / \sum_{j=1}^{m} s_{xj}$	The mean fecundity of individuals at age x [40]
Net reproductive rate (R_0) and cumulative net reproductive rate (R_x)	$R_0 = \sum_{x=0}^{\infty} l_x m_x, R_x = \sum_{i=0}^{x} l_i m_i$	R_0 is the total offspring that an average individual (including females, males, and those that died in the pre-adult stage) can produce during its lifetime [40]. R_x is the total offspring that an average individual can produce from age 0 to age x.
Intrinsic rate of increase (r)	$\sum_{x=0}^{\infty} e^{-r(x+1)} l_x m_x = 1$	The population growth rate as time approaches infinity and the population reaches the stable age-stage distribution (SASD). It is calculated by using the Euler–Lotka equation with age indexed from 0 [40].
The finite rate of increase (λ)	$\lambda = e^r$	The population growth rate as time approaches infinity and the population reaches a stable age-stage distribution. The population size will increase at the rate of λ per unit time [40].
Mean generation time (T)	$T = \ln R_0 / r$	The time length that a population requires to increase to R_0-fold of its size as time approaches infinity and the population settles down to a stable age-stage distribution [40].

Table A2. *Cont.*

Parameter	Equation	Definition and Reference
Age-stage life expectancy (e_{xj})	$e_{xj} = \sum_{i=x}^{\infty} \sum_{y=j}^{m} s'_{iy}$	The time length that an individual of age x and stage j is expected to live. The notation s'_{iy} is the probability that an individual of age x and stage j will survive to age i and stage y. It is calculated according to Chi and Liu (1985) by assuming s_{xj} = 1 [26,40].
Age-stage reproductive value (v_{xj})	$v_{xj} = \frac{e^{r(x+1)}}{s_{xj}} \sum_{i=x}^{\infty} e^{-r(x+1)} \sum_{y=j}^{m} s'_{iy} f_{iy}$	The contribution of an individual of age x and stage j to the future population [27].

References

1. Wang, R.L.; Jiang, C.X.; Guo, X.; Chen, D.D.; You, C.; Zhang, Y.; Wang, M.T.; Li, Q. Potential distribution of *Spodoptera frugiperda* (J.E. Smith) in China and the major factors influencing distribution. *Glob. Ecol. Conserv.* **2020**, *21*, e00865. [CrossRef]
2. Zhou, Y.; Wu, Q.L.; Zhang, H.W.; Wu, K.M. Spread of invasive migratory pest *Spodoptera frugiperda* and management practices throughout China. *J. Integr. Agric.* **2021**, *20*, 637–645. [CrossRef]
3. Yang, F.; Williams, J.; Huang, F.N.; Kerns, D.L. Genetic basis and cross-resistance of Vip3Aa resistance in *Spodoptera frugiperda* (Lepidoptera: Noctuidae) derived from Texas, USA. *Crop Prot.* **2021**, *147*, 105702. [CrossRef]
4. Li, X.J.; Wu, M.F.; Ma, J.; Gao, B.Y.; Wu, Q.L.; Chen, A.D.; Liu, J.; Jiang, Y.Y.; Zhai, B.P.; Early, R.; et al. Prediction of migratory routes of the invasive fall armyworm in eastern China using a trajectory analytical approach. *Pest Manag. Sci.* **2020**, *76*, 454–463. [CrossRef] [PubMed]
5. Du Plessis, H.; Schlemmer, M.-L.; Van Den Berg, J. The effect of temperature on the development of *Spodoptera frugiperda* (Lepidoptera: Noctuidae). *Insects* **2020**, *11*, 228. [CrossRef]
6. Montezano, D.G.; Specht, A.; Sosa-Gómez, D.R.; Roque-Specht, V.F.; Sousa-Silva, J.C.; Paula-Moraes, S.V.; Peterson, J.A.; Hunt, T.E. Host plants of *Spodoptera frugiperda* (Lepidoptera: Noctuidae) in the Americas. *Afr. Entomol.* **2018**, *26*, 286–300. [CrossRef]
7. Nagoshi, R.N.; Meagher, R.L. Behavior and distribution of the two fall armyworm host strains in Florida. *Fla. Entomol.* **2004**, *87*, 440–449. [CrossRef]
8. Dumas, P.; Legeai, F.; Lemaitre, C.; Scaon, E.; Orsucci, M.; Labadie, K.; Gimenez, S.; Clamens, A.-L.; Henri, H.; Vavre, F.; et al. *Spodoptera frugiperda* (Lepidoptera: Noctuidae) host-plant variants: Two host strains or two distinct species? *Genetica* **2015**, *143*, 305–316. [CrossRef]
9. Zhang, L.; Jin, M.H.; Zhang, D.D.; Jiang, Y.Y.; Liu, J.; Wu, K.M.; Xiao, Y.T. Molecular identification of invasive fall armyworm *Spodoptera frugiperda* in Yunnan Province. *Plant Prot.* **2019**, *45*, 19–24.
10. Xu, C.F.; Sa, R.N.; He, J.Y.; Guo, X.; Lu, F.; Cheng, W.; Hu, G.; Chen, F.J.; Wan, G.J.; Zhao, J.Y. Comparative study on the adaptability of *Spodoptera frugiperda* to different corn varieties. *J. Environ. Entomol.* **2021**, *43*, 891–900.
11. Wan, J.; Huang, C.; Li, C.Y.; Zhou, H.X.; Ren, Y.L.; Li, Z.Y.; Xing, L.S.; Zhang, B.; Qiao, X.; Liu, B.; et al. Biology, invasion and management of the agricultural invader: Fall armyworm, *Spodoptera frugiperda* (Lepidoptera: Noctuidae). *J. Integr. Agric.* **2021**, *20*, 646–663. [CrossRef]
12. Kenis, M.; Benelli, G.; Biondi, A.; Calatayud, P.A.; Day, R.; Desneux, N.; Harrison, R.D.; Kriticos, D.; Rwomushana, I.; Van Den Berg, J.; et al. Invasiveness, biology, ecology, and management of the fall armyworm, *Spodoptera frugiperda*. *Entomol. Gen.* **2023**, *43*, 187–241. [CrossRef]
13. Ning, S.Y., Yang, H.Y.; Fan, D.S.; Feng, J.N. Influence of larval experience on preference of a subterranean insect *Delia Antiqua* on *Allium* hosts. *J. Appli. Entomol.* **2018**, *142*, 263–271. [CrossRef]
14. Hopkins, A.D. A discussion of C. G. Hewitt's paper on "Insect Behaviour". *J. Econ. Entomol.* **1917**, *10*, 92–93.
15. Liao, S.K.; Lin, H.Y.; Wang, J.J.; Wang, Q.; Wei, H.J.; Chen, H. Effects of different *Ficus* feeding experiences on host preference of *Perina nuda* larvae (Lepidoptera: Lymantriidae). *J. Econ. Entomol.* **2024**, *117*, 209–217. [CrossRef]
16. Videla, M.; Valladares, G.; Salvo, A. Differential effects of experience on feeding and ovipositing preferences of a polyphagous leafminer. *Entomol. Exp. Appli.* **2010**, *137*, 184–192. [CrossRef]
17. Boiça Júnior, A.L.; Souza, B.H.S.; Costa, E.N.; Paiva, L.B. Influence of fall armyworm previous experience with soybean genotypes on larval feeding behavior. *Arth. Plant Interact.* **2017**, *11*, 89–97. [CrossRef]

18. Acharya, R.; Malekera, M.J.; Dhungana, S.K.; Sharma, S.R.; Lee, K.-Y. Impact of rice and potato host plants is higher on the reproduction than growth of corn strain fall armyworm, *Spodoptera frugiperda* (Lepidoptera: Noctuidae). *Insects* **2022**, *13*, 256. [CrossRef]
19. Li, D.Y.; Zhi, J.R.; Zhang, T.; Ye, J.Q.; Liang, Y.J. Effects of different host plants on the development and reproduction of *Spodoptera frugiperda*. *J. Environ. Entomol.* **2020**, *42*, 311–317.
20. Lu, J.J.; Zhang, B.; Zhuang, M.M.; Ren, M.F.; Li, D.Q.; Yan, H.W.; Long, J.M.; Jiang, X.L. Preference and performance of the fall armyworm, *Spodoptera frugiperda*, on six cereal crop species. *Entomol. Exp. Appl.* **2023**, *171*, 492–501. [CrossRef]
21. Wu, G.; Su, L.; Li, J.S.; Xiao, N.W.; Chen, F.J.; Xiong, Y.F. Effects of plant chemicals on the cotton bollworm, *Helicoverpa armigera* (Hübner): A multi-generation study. *Int. J. Pest Manag.* **2012**, *58*, 361–368. [CrossRef]
22. Ma, L.; Wang, D.T.; Ren, Q.L.; Zhang, L.; Cheng, Y.X.; Jiang, X.F. Feeding experience on hairy vetch induces a loss of host-specific adaptation to maize and changes in the gut microbiota communities of the fall armyworm. *J. Pest Sci.* **2024**, *97*, 777–792. [CrossRef]
23. Chi, H.; You, M.S.; Atlıhan, R.; Smith, C.L.; Kavousi, A.; Özgökçe, M.S.; Güncan, A.; Tuan, S.J.; Fu, J.W.; Xu, Y.Y.; et al. Age-stage, two-sex life table: An introduction to theory, data analysis, and application. *Entomol. Gen.* **2020**, *40*, 103–124. [CrossRef]
24. Yang, Y.J.; Xu, H.X.; Hu, Y.; Han, H.L.; Qian, J.N.; Yin, C.; Lü, Z.X. Growth development and reproduction of *Spodoptera frugiperda* reared on an artificial diet. Chin. *J. Appl. Entomol.* **2020**, *57*, 1341–1344.
25. Chi, H.; Güncan, A.; Kavousi, A.; Gharakhani, G.; Atlihan, R.; Özgökçe, M.S.; Shirazi, J.; Amir-Maafi, M.; Maroufpoor, M.; Roya, T. TWOSEX-MSChart: The key tool for life table research and education. *Entomologia* **2022**, *42*, 845–849. [CrossRef]
26. Chi, H.; Su, H.Y. Age-stage, two-sex life tables of *Aphidius gifuensis* (Ashmead) (Hymenoptera: Braconidae) and its host *Myzus persicae* (Sulzer) (Homoptera: Aphididae) with mathematical proof of the relationship between female fecundity and the net reproductive rate. *Environ. Entomol.* **2006**, *35*, 10–21. [CrossRef]
27. Tuan, S.J.; Lee, C.C.; Chi, H. Population and damage projection of *Spodoptera litura* (F.) on peanuts (*Arachis hypogaea* L.) under different conditions using the age-stage, two-sex life table. *Pest Manag. Sci.* **2014**, *70*, 805–813. [CrossRef]
28. Rousselet, G.A.; Pernet, C.R.; Wilcox, R.R. An introduction to the bootstrap: A versatile method to make inferences by using data-driven simulations. *Entomol. Gen.* **2020**, *40*, 103–124. [CrossRef]
29. Yu, C.D.; Huang, J.; Ren, X.Y.; Fernández-Grandon, G.M.; Li, X.Y.; Hafeez, M.; Lu, Y.B. The predatory bug *Orius strigicollis* shows a preference for egg-laying sites based on plant topography. *PeerJ* **2021**, *9*, e11818. [CrossRef]
30. Zhu, J.Y.; Xiang, Z.W.; Zhang, S.Z.; Wu, L.N.; Liu, T.X. Adaptations of *Plutella xylostella* adult females and larvae to waxy host plants. *J. Pest Sci.* **2022**, *95*, 203–214. [CrossRef]
31. Nataraj, N.; Adam, E.; Hansson, B.S.; Knaden, M. Host plant constancy in ovipositing *Manduca sexta*. *J. Chem. Ecol.* **2021**, *47*, 1042–1048. [CrossRef] [PubMed]
32. Awmack, C.S.; Leather, S.R. Host plant quality and fecundity in herbivorous insects. *Annu. Rev. Entomol.* **2002**, *47*, 817–844. [CrossRef]
33. Wang, Z.; Zhu, H.; Huang, J.; Jin, D.C.; Yang, H. Feeding preference and fitness of *Spodoptera frugiperda* on rice and their relationships with biochemical substance contents of leaves. *Plant Prot.* **2023**, *49*, 21–28.
34. Dai, Q.X.; Li, Z.Y.; Tian, Y.J.; Zhang, Z.F.; Wang, L.; Lu, Y.Y.; Li, Y.Z.; Chen, K.W. Effects of different corn varieties on development and reproduction of *Spodoptera frugiperda*. *Chin. J. Appl. Ecol.* **2020**, *31*, 3273–3281.
35. Zhang, Q.Y.; Zhang, Y.L.; Quandahor, P.; Gou, Y.P.; Li, C.C.; Zhang, K.X.; Liu, C.Z. Oviposition preference and age-stage, two-sex life table analysis of *Spodoptera frugiperda* (Lepidoptera: Noctuidae) on different maize varieties. *Insects* **2023**, *14*, 413. [CrossRef]
36. Siscaro, G.; Lo Pumo, C.; Tropea Garzia, G.; Tortorici, S.; Gugliuzzo, A.; Ricupero, M.; Biondi, A.; Zappalà, L. Temperature and tomato variety influence the development and the plant damage induced by the zoophytophagous mirid bug *Nesidiocoris tenuis*. *J. Pest Sci.* **2019**, *92*, 1049–1056. [CrossRef]
37. Lü, L.; Li, Y.Q.; Chen, C.L.; Chang, X.Q.; Zhang, S.; Xu, D.; Wan, P. Comparison of the feeding and growth characteristics of larvae of the fall armyworm, *Spodoptera frugiperda* (Lepidoptera:Noctuidae), on corn and wheat. *Acta Entomol. Sin.* **2020**, *63*, 597–603. [CrossRef]
38. Soto, E.M.; Padró, J.; Milla Carmona, P.; Tuero, D.T.; Carreira, V.P.; Soto, I.M. Pupal emergence pattern in cactophilic *Drosophila* and the effect of host plants. *Insect. Sci.* **2018**, *25*, 1108–1118. [CrossRef]
39. Wang, T.; Xu, H.X.; Yang, Y.J.; Zheng, X.S.; Lü, Z.X. Effects of different rice varieties on the growth and development fecundity of *Cnaphalocrocis medinalis* Guenée. *J. Environ. Entomol.* **2018**, *40*, 532–535.
40. Chi, H.; Liu, H. Two new methods for the study of insect population ecology. *Bull. Inst. Zool. Acad. Sin.* **1985**, *24*, 225–240.
41. Lu, Z.H.; Chen, Y.P.; Zhou, A.C.; He, S.Q.; Li, H.; Bao, Y.Y.; Gui, F.R. Effects of multi-generation feeding with different host plants on activities of enzyme in *Spodoptera frugiperda* larvae. *J. Environ. Entomol.* **2020**, *42*, 1361–1368.
42. Guo, J.Y.; Wu, G.; Wan, F.H. Activities of digestive and detoxification enzymes in multiple generations of beet armyworm, *Spodoptera exigua* (Hübner), in response to transgenic Bt cotton. *J. Pest Sci.* **2010**, *83*, 453–460. [CrossRef]

43. Zhang, Y.K.; Zhu, G.Y.; Chen, B.; Duan, B.; Zhang, Z.B.; Li, Z.Y.; Xiao, G.L. Effect of successive rearing on the developmental duration of *Bactrocera correcta*. *J. Yunnan Agric. Univ.* **2015**, *30*, 676–680.
44. Du, X.Y.; Li, H.S.; Chen, P.T.; Pang, H. Alternative foods for the rearing of a ladybird: Successive generations of the *Micraspis discolor* fed on *Ephestia kuehniella* eggs and *Brassica campestris* pollen. *J. Environ. Entomol.* **2021**, *43*, 305–314.
45. Lu, J.J.; Zhuang, M.M.; Long, J.M.; Jiang, X.L.; Zhang, Y.X.; Ren, M.F.; Li, D.Q.; Zhang, B.; Wu, Y.P.; Li, G.P.; et al. Linking life table and consumption rate of the fall armyworm, *Spodoptera frugiperda* reared on different maize cultivars. *Entomo. Gen.* **2024**, *44*, 961–969. [CrossRef]

insects

MDPI

Article

Males of *Dalbulus maidis* Attract Females Through Volatile Compounds with Potential Pheromone Function: A Tool for Pest Management

Mateus Souza Sanches [1,2,*], Miguel Borges [1], Raul Alberto Laumann [1], Charles Martins Oliveira [3], Marina Regina Frizzas [2] and Maria Carolina Blassioli-Moraes [1,*]

1 Laboratório de Semioquímicos, Embrapa Recursos Genéticos e Biotecnologia, Brasília, DF 70770-917, Brazil; miguel.borges@embrapa.br (M.B.); raul.laumann@embrapa.br (R.A.L.)
2 Programa de Pós-Graduação em Ecologia, Universidade de Brasília, Brasília, DF 70910-900, Brazil; frizzas@unb.br
3 Embrapa Cerrados, Planaltina, DF 73310-970, Brazil; charles.oliveira@embrapa.br
* Correspondence: mateus.sanches767@gmail.com (M.S.S.); carolina.blassioli@embrapa.br (M.C.B.-M.)

Simple Summary

Insects use chemical compounds for communication, and sex pheromone is one of the most important signals used by males and females to find each other for mating purposes. The corn leafhopper, *Dalbulus maidis*, is an insect vector that transmits pathogens causing diseases in maize crops, but it was unknown whether it uses sex pheromones in their communication. In this study, we tested whether *D. maidis* produces volatile compounds that attract the opposite sex. We collected volatiles from live insects and evaluated their influence on the behavioral responses of conspecifics. We found that males produce odors that attract females. Interestingly, males avoided odors emitted by stressed females, which may suggest the release of an alarm pheromone. These findings highlight for the first time the role of semiochemicals in intraspecific *D. maidis* communication, and open new perspectives for the development of monitoring and management tools targeting this important pest.

Abstract

Insects use chemical compounds to communicate with conspecifics and other organisms. The corn leafhopper, *Dalbulus maidis* (Hemiptera: Cicadellidae) (DeLong & Wolcott), is an important pest in Brazilian maize crops due to its role as a vector of phytopathogens. Despite its economic importance, the chemical communication between sexes in this species remains to be elucidated. This research aimed to unveil whether *D. maidis* produces chemical compounds that influence the behavior of the opposite sex and may act as sex pheromones. To evaluate the influence of these volatiles, olfactometer bioassays were conducted as dynamic headspace volatile collections from live insects. Results showed that both male and female leafhoppers emit volatile compounds; however, no sex-specific compounds were detected. Females were attracted to male odors and male aeration extracts, suggesting males produce sex-specific volatiles. Interestingly, males avoided odors from non-acclimated females, which may indicate possible alarm pheromone release. Although the compounds were not identified, this is the first study to demonstrate intraspecific chemical communication in *D. maidis* mediated by volatiles, and the first such record in Membracoidea. These results contribute to understanding the pest's biology and support the development of monitoring and control strategies in maize crops.

Keywords: semiochemicals; chemical communication; olfactometry; corn leafhopper; pest management; Hemiptera; Cicadellidae

1. Introduction

Insects rely on chemical communication to interact with conspecifics (pheromones) or with organisms from other species (allelochemicals) [1,2]. Sex pheromones are chemical compounds released by sexually mature individuals that function to stimulate the opposite sex for mate location and copulation. These pheromones can also convey information about mate recognition, attraction, reproductive status, and the fitness of the emitter [3].

The corn leafhopper, *Dalbulus maidis* (DeLong & Wolcott), is a phloem-feeding insect that primarily causes indirect damage to maize *Zea mays* L., its main host plant [4], due to its ability to efficiently transmit the pathogens associated with the corn stunt disease complex. These include corn stunt spiroplasma (CSS, *Spiroplasma kunkelii*), maize bushy stunt phytoplasma (MBSP, '*Candidatus* Phytoplasma asteris'), as well as maize rayado fino virus (MRFV) and maize striate mosaic virus (MSMV) [4–6]. Currently, no curative management measures are available for these diseases [7,8].

The mating behavior of *D. maidis* has been previously described, including the use of vibrational and acoustic signals during courtship [9,10]. However, no studies have investigated whether this species emits long-range chemical signals, like sex pheromones, considering that the acoustic signals produced by *D. maidis* are transmitted only over short distances when both sexes are on the same substrate [9]. Chemical communication is, nevertheless, known to play a role in host plant selection by *D. maidis* [11,12].

Although several studies have demonstrated the use of acoustic signals among leafhoppers, for a long time, there was no evidence of molecules functioning as sex pheromones in any species within the group (Hemiptera: Auchenorrhyncha: Cicadomorpha). This scenario began to change about a decade ago with the identification of an aggregation pheromone in *Callitettix versicolor* (Fabricius) [13], and, more recently, the first evidence of sex pheromones was reported for *Philaenus spumarius* (L.), although the chemical structure has not yet been identified [14].

Due to their high specificity and the remarkable sensitivity of insect olfactory systems, sex pheromones have been widely studied as tools for agricultural pest management [15,16]. Since it is a vector insect, there is no established economic threshold for *D. maidis*, and its management still relies primarily on the systematic use of chemical insecticides during the early stages of maize cultivation, which is the critical period for pathogen transmission [12,17]. Therefore, alternative control strategies are urgently needed, including the potential use of sex pheromones for monitoring and management.

In this context, the objective of this study was to investigate whether *D. maidis* emits volatile compounds that can influence the behavioral response of conspecifics, with a particular focus on the presence of a sex pheromone.

2. Materials and Methods

2.1. Corn Leafhopper-Dalbulus Maidis

Dalbulus maidis (DM) used in this study were obtained from a colony established in 2022 at Embrapa Cerrados (Planaltina, DF, Brazil), originally collected from adult individuals in experimental maize fields (15°36′16″S, 47°42′38″W). The colony was subsequently decontaminated following the recommended procedures [18] to eliminate potential carriers of pathogens (mollicutes and viruses), ensuring a healthy colony.

The individuals used in the experiments were randomly selected from adult insects, without controlling for age, reproductive status, or mating history. To determine the sex, insects were placed in glass tubes and examined under a stereomicroscope. Females were identified by the presence of an ovipositor at the tip of the abdomen (Figure S1).

2.2. Maize

Maize plants of the Synthetic Spodoptera (SS) genotype were used in the experiments since their herbivore-induced plant volatiles are well known [19]. The seeds were obtained from the germplasm bank of Embrapa Maize and Sorghum (Sete Lagoas, MG, Brazil) and were sown in plastic pots (0.3 L) with a mixture of natural soil (Latosol) and organic substrate (Max Fertil, Santa Catarina, Brazil, composition: pine bark, natural phosphate, carbonized rice husk, vermiculite, chemical fertilizer NPK) in a proportion of 1:1 w/w, without post-fertilization. Plants were kept in a greenhouse (Brasília, DF, Brazil) under natural conditions of temperature, humidity, and photoperiod (14L/10D; Brasília, DF, 15°46′46″ S and 47°55′46 W) and manually irrigated with a watering can every two days. For experiments, plants were used at the V2 stage (two expanded leaves, 4 to 7 days after germination).

2.3. Volatile Collection

Volatile organic compounds (VOCs) emitted by *D. maidis* were collected using dynamic headspace aeration systems across two experimental sets that lasted for 2 weeks each. In the first experimental set three treatments were established: DM: 100 unsexed adult leafhoppers; DM-Maize: 100 unsexed adults plus one maize plant at V2 growth stage, and Maize: a maize plant at the V2 stage. The objectives were as follows: DM to assess volatiles emitted exclusively by the leafhoppers; DM-Maize to evaluate whether volatiles are only emitted in the presence of a food resource; Maize to identify and exclude volatiles released by the plant itself. For each treatment 20 replicates were conducted. The maize plants were placed intact in the system, with aluminum covering the soil to prevent volatiles from originating from the soil or roots (Figure S2).

In the second experimental set, two other treatments were conducted: volatiles were collected from 100 males or 100 females and another with one maize plant at the V2 growth stage, these treatments were called: DM-Male (100 males) and DM-Female (100 females). For each treatment 10 replicates were conducted.

All treatments were conditioned in glass chambers (2L volume; 14 cm × 24 cm) connected to a system supplied with charcoal-filtered air (4–20 mesh, Supelco, Pennsylvania, USA) at a constant flow rate of 1.0 L min^{-1}. Simultaneously, a vacuum pump (LGI-DVP-1, ultimate vacuum 200 mbar, pumping speed 60 L/min, LGI Scientific, São Paulo, Brazil) maintained an outflow of 0.8 L min^{-1}, ensuring continuous air circulation through the chamber and directing volatiles into an adsorbent tube filled with HayeSep Q (polydivinylbenzene copolymer, 100 mg, 80–100 mesh, Supelco, PA, USA).

Volatiles trapped on adsorbent filters were eluted every 24 h with n-hexane, with 72 h accumulations during weekends. All insects and plants were replaced weekly with new ones. For the DM treatment, which did not include maize plants, six replicates were conducted, and with volatiles collected for 24 h, due to high mortality (individuals typically died after 24 h without food).

2.4. Chemical Analysis

All volatile samples were analyzed using an Agilent 7890A gas chromatograph equipped with a flame ionization detector (GC-FID) and a non-polar DB-5MS column

(30 m × 0.25 mm ID, 0.25 μm film thickness, Supelco, PA, USA). The oven temperature was initially set at 40 °C for 2 min, then ramped at 5 °C min^{-1} until reaching 180 °C, held for 0.1 min, followed by an increase of 10 °C min^{-1} to 250 °C, where it was held for 20 min. The injector was set to 250 °C and the FID to 270 °C. Each sample (2 μL) was injected in splitless mode, with helium as the carrier gas. Chromatographic data acquisition was performed using the GC ChemStation software (Agilent, Santa Clara, CA, USA, version 2.4). For qualitative analyses selected volatile samples was analyzed using an Agilent 5975 mass selective detector (GC-MS) (Agilent, Santa Clara, CA, USA) coupled with a quadrupole mass analyzer, equipped with the same non-polar DB-5MS capillary column (30 m × 0.25 mm ID, 0.25 μm film thickness; J&W Scientific, Folsom, CA, USA). Injections were performed in splitless mode with 2 μL of sample, and helium was used as the carrier gas. Ionization was conducted by electron impact (EI) at 70 eV, with the ion source temperature set at 230 °C. The oven program followed the same temperature profile as the GC-FID analyses: 40 °C for 2 min, ramped at 5 °C min^{-1} to 180 °C, followed by 10 °C min^{-1} to 250 °C, with a final hold of 20 min. Data acquisition and analysis were conducted using the MassHunter Qualitative software (version 10.1, Agilent, Santa Clara, CA, USA). Compound identification was performed by comparing the mass spectra to those in the NIST library [20] and to published spectra, along with retention index calculations based on the DB-5MS column. Tentative identifications were confirmed by co-injection with authentic standards, either commercially sourced or synthesized in-house.

2.5. Chemicals

Authentic chemical standards of decane (99%), tetradecane (99%), pentadecane and hexadecane (99), octanal (≥98.0% (GC)), nonanal (98%), decanal (98%), dodecanal (97%), camphene (95%), α-pinene (98%), β-pinene (99%), 3-carene (≥90%), limonene (97%), methyl salicylate (99%), 6-methyl-5-hepten-2-one (99%), β-caryophyllene (98%), geranylacetone (97%) and cyclosativene (99%) were purchased from Sigma-Aldrich (Steinheim, Germany). Linalool was purchased from TCI America (Portland, USA). (*E*)-4,8-Dimethylnona-1,3,7-triene (DMNT) (95%) was synthesized from geraniol [21]. The solvent hexane (97%) was purchased from Sigma-Aldrich (Steinheim, Germany) and redistilled before use.

2.6. Olfactometer Bioassays

To assess whether *D. maidis* is attracted to the opposite sex through chemical cues, olfactometer bioassays were conducted using a Y-shaped dual-choice olfactometer (19 cm × 19 cm; choice arms: 7 cm × 1.7 cm; main arm: 5 cm × 1.7 cm, Acrilico Arte, Brasília, DF, Brazil). Before the experiments with the odor sources, the behavioral response of insects of both sexes were evaluated when exposed to clean air to assess any potential bias that might affect their choice. Then, female and male responses were tested separately by contrasting the odor of 20 live males or 20 live females and clean air in the following pairs: air–air, male–female, male–air, and female–air. Odor sources were generated by placing the insects inside a 40 mL glass container (5 cm × 6 cm) without food. The insects were transferred to the experimental arenas using an insect mouth aspirator [18], from the rearing cages to glass tubes, and, subsequently, to the olfactometers.

Initially, insects were placed into the containers and bioassays were immediately started in the afternoon (n = 30). One insect (male or female) was released at the entrance of the olfactometer, and its behavior was monitored for 10 min. Because the odor source insects in the glass container were highly mobile, bioassays were repeated using acclimated individuals. Acclimation was achieved by keeping the insects intended as odor sources inside the container for two hours. After this period, the odor sources were connected to the olfactometer, resulting in clearer response behaviors from the test insects toward the odors being evaluated.

Since only females had been attracted to the odor of live males in the previous experiment, they were tested with a mixture (10 females and 10 males together) versus males, as well as with male aeration extracts (obtained from 24 h volatile collections, DM-Male) against n-hexane. For male aeration extract bioassays, 5 μL of DM-Male samples (equivalent to the aeration of 20 individuals) were applied to a paper filter (0.5 cm × 0.5 cm) and compared to 5 μL of n-hexane. The paper filters were placed in glass containers (20 mL; 3 cm × 3 cm) and connected to the system. After three bioassays (10 min each), the filter papers containing the 5 μL of the sample, volatile or hexane, were replaced with new ones.

Thirty different individuals of the tested sex were used to evaluate their choice in response to the odor treatments. The parameters measured were first choice (the first arm entered for more than 30 s) and residence time (total time spent in each arm). Individuals that made no choice within this period were classified as non-responsive and excluded until 30 bioassays with a choice were obtained. To avoid contamination with chemical cues, the olfactometer was cleaned after every five bioassays, and odor source positions were alternated to prevent side bias. A maximum of 10 bioassays were conducted per day to minimize the influence of environmental factors such as temperature, humidity, atmospheric pressure and insect condition. All bioassays were performed between 14:00 and 17:00 h.

2.7. Statistical Analyses

Olfactometers bioassays data were analyzed using a chi-square test for first choice and paired t-tests for residence time with a 95% level of confidence; all tests were conducted on the R platform [22].

3. Results

3.1. Bioassays

Insects of both sexes exhibited no directional preference when exposed solely to clean air on both sides in olfactometers (females: $\chi^2 = 0.31$, $p = 0.57$; t-test = 0.41, $p = 0.2$; males: $\chi^2 = 0.043$, $p = 0.83$; t-test = 0.55, $p = 0.58$).

Females discriminated between males and females, being attracted to males ($\chi^2 = 6.533$, $p = 0.01$; t-test = 3.03, $p = 0.005$), however, they did not differentiate males from the mixture of males and females ($\chi^2 = 0.533$, $p = 0.46$; t-test = 1.69, $p = 0.1$) (Figure 1A and B). When exposed to the odor of non-acclimated males, *D. maidis* females did not show a preference for male odor compared to air in terms of first choice ($\chi^2 = 3.33$, $p = 0.067$), but they spent significantly more time in the arm containing male odor (t-test = 25.299, $p = 0.017$) (Figure 1A and B). Using the odor of acclimated males, females corroborated the previous results with non-acclimated males, and showed a clear preference for male odor, both in first choice ($\chi^2 = 10.8$, $p = 0.001$) and in residence time (t-test = 33.104, $p = 0.002$) (Figure 1A and B). Females also showed a preference to male aeration extract compared to hexane ($\chi^2 = 8.533$, $p = 0.003$; t-test = 3.023, $p = 0.005$). Finally, females did not show a significant preference for female odor compared to air ($\chi^2 = 0.142$, $p = 0.705$; t-test = -1.145, $p = 0.261$).

Males exhibited avoidance behavior toward female odor when females were not acclimated for a two-hour period (Figure 2A and B). They showed a significant preference for the arm releasing clean air over the odor of 20 females, both in first choice ($\chi^2 = 4.8$, $p = 0.024$) and residence time (t-test = 22.838, $p = 0.03$). On the other hand, when females were acclimated, males showed no attraction either to female odor or to clean air, displaying no significant preference ($\chi^2 = 0.133$, $p = 0.718$ for first choice; t-test = 0.972, p

= 0.338 for residence time). They did not discriminate between males and females (χ^2 = 0.133, p = 0.715; t-test = –0.09, p = 0.924), nor when exposed to male odor, showing no side preference (χ^2 = 0.615, p = 0.432; t-test = –0.267, p = 0.791).

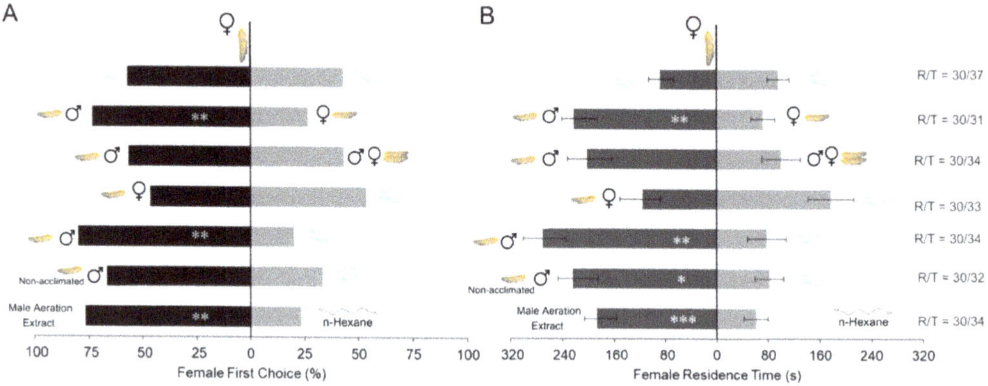

Figure 1. (**A**) First choice of *Dalbulus maidis* females in dual-choice olfactometer bioassays between different odors sources: 20 males, 20 females, 10 males and 10 females, clean air, male aeration extract and n-hexane (n = 30). (**B**) Mean residence time of females during the bioassays. * indicates $p < 0.05$, ** $p < 0.01$ and *** $p < 0.001$ for chi-square tests (A) and t-tests (B). Error bars represent the standard error of the mean. R/T indicates the number of responsive/total number of bioassays conducted.

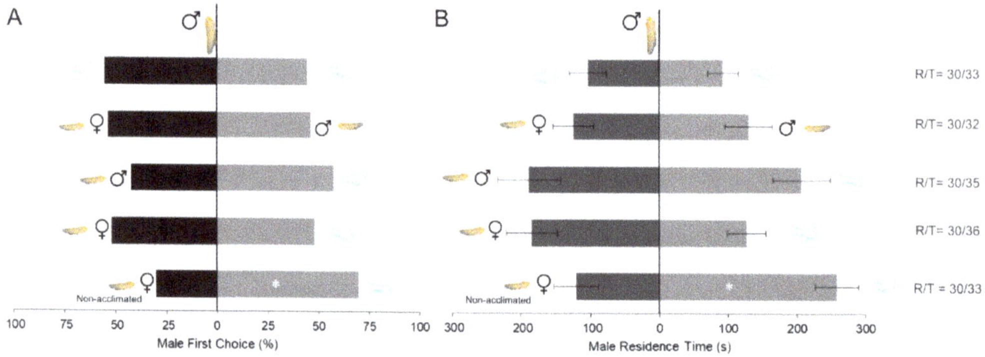

Figure 2. (**A**) First choice of *Dalbulus maidis* males in dual-choice olfactometer bioassays between different odor sources: 20 males, 20 females and clean air (n = 30). (**B**) Mean residence time of males during the bioassays. * indicates $p < 0.05$ for chi-square tests (**A**) and t-tests (**B**). Error bars represent the standard error of the mean. R/T indicates the number of responsive/total number of bioassays conducted.

3.2. Volatiles

A total of 30 volatile compounds were identified from headspace collections containing either male or female *D. maidis* individuals that were similar for both sexes (Figure 3). The chemical profile of volatiles obtained when only the insects were used for volatile collection, did not show a qualitative difference between both sexes. The main compounds identified were monoterpenes like α (2) and β (4) pinenes, limonene and linalool (14), sesquiterpenes like β-caryophyllene and aldehydes (Figure 3). The chemical analysis of volatile samples conducted with both sexes did not show differences compared with when the insects had their volatiles collected and separated by sex. When maize plants were used as food sources for the insects during the volatile collections, we did not detect different compounds that could be attributed to the insects. The new peaks observed in the chromatographic profile in the presence of both insects and maize plants were associated with plant-derived compounds, such as DMNT (16) and cyclosativene (26) (Figure 3).

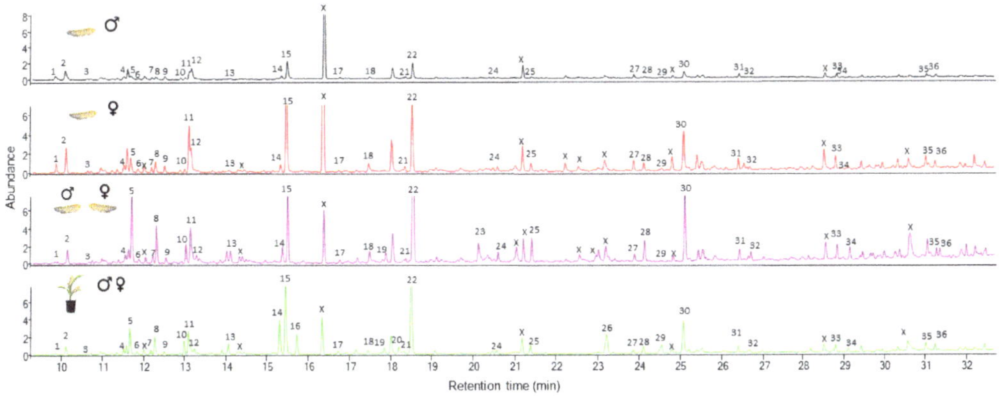

Figure 3. Chromatograms of headspace collections from males, females, both sexes and both sexes with maize plants analyzed by gas chromatography coupled with mass spectrometry (GC-MS). X indicates phthalate group contaminants. 1–β-thujone, 2–α-pinene, 3–camphene, 4–β-pinene, 5–6-methyl-5-hepten-2-one, 6–2-pentylfuran, 7–decane, 8–octanal, 9–3-carene, 10–p-cymene, 11–2-ethyl-1-hexanol, 12–limonene, 13–γ-terpinene, 14–linalool, 15–nonanal, 16–DMNT: (E)-4,8-dimethyl-1,3,7-nonatriene, 17–2-ethylhexyl acetate, 18–NI-1, 19–terpinen-4-ol, 20–methyl salicylate, 21–dodecane, 22–decanal, 23–4-phenyl-2-butanol, 24–NI-2, 25–NI-3, 26–cyclosativene, 27–tetradecane, 28–dodecanal, 29–(E)-β-caryophyllene, 30–geranylacetone, 31–pentadecane, 32–tridecanal, 33–hexadecane, 34–tetradecanal, 35–heptadecane, 36–pentadecanal.. NI = non-identified.

4. Discussion

The bioassay results showed that female corn leafhoppers, *D. maidis*, can discriminate between males and females through volatiles, moving toward the odor of live males and male aeration extracts, thereby suggesting that males produce volatiles functioning as a pheromone. Males were not attracted to these conspecific volatiles of both sexes, as the females did not respond to the odor of other females, supporting the hypothesis that these compounds are only released by males and act as sexual attractants exclusively for females. The chemical analysis of volatile samples containing the volatiles emitted by *D. maidis* did not identify male-specific compounds that could explain this attraction. It is possible that the male-specific compounds are produced in very tiny amounts, below the detection limits of the equipment used. While the behavioral assays strongly suggest a role in sexual attraction, the precise function as a sex pheromone remains to be confirmed until the active compound is chemically identified and the behavioral role of the synthetic pheromone is evaluated.

In both conditions used to collect volatiles from males and females, the main compounds identified were aldehydes, monoterpenes, and linear hydrocarbons. No sex-specific compounds were observed between the volatile profiles of males and females collected without food and those of males and females collected with food. In samples with and without food we mainly found compounds related to maize plant emissions, such as cyclosativene and DMNT. The presence of these compounds, even in the absence of maize plants, may be related to honeydew excretion by the leafhoppers, which could not be controlled under our methodology. In *Lycorma delicatula* (White) (Fulgoridae), honeydew volatiles released by males can attract conspecific males [23]; therefore, this may represent a form of chemical communication in Auchenorrhyncha insects.

The absence of food can influence pheromone production, as demonstrated by [24]. The stink bug *Euschistus heros* (Fabricius) ceases sex pheromone production after remaining more than 24 h without food, before this period males of *E. heros* cotinue producing the sex pheromone. In the present study, we were not able to identify sex pheromone compounds and, therefore, cannot determine precisely whether the absence of food affects or not pheromone emission. All insects used in the experiments were fed until the start of the assays, remained without food for only 24 h; thus, we hypothesized that at least during the first hours of the aeration, the insects had physiological conditions to produce and emit semiochemicals.

Male of leafhoppers mate multiple times throughout their lives [9]. Therefore, it is reasonable to expect that mated individuals may continue producing and emitting sex pheromones. The results obtained in this study appear to support this hypothesis, as part of the insects used were mated and males were still able to attract females in olfactometer bioassays, even though females mate only once in their lives [9]. Further research should be conducted using virgin males and females to assess whether virgin individuals exhibit stronger behavioral responses in bioassays and whether they produce higher levels of sex pheromones. However, this attraction, even in mated females, should be considered an advantage for potential control methods in the field.

Mating communication in leafhoppers (Auchenorrhyncha) has been characterized primarily by acoustic and vibrational signals [25–27]. However, multiple studies have demonstrated the use of chemical cues in these insects for host-plant location [28,29], including in *D. maidis* [11,12]. Only recently has evidence emerged for the presence of sex pheromones within this suborder. In 2022, the first indication of a sex pheromone was reported for *P. spumarius* (Hemiptera: Auchenorrhyncha: Aphrophoridae), where males were attracted to female odors in Y-tube olfactometer assays [14]. Similarly, *L. delicatula* males were shown to be attracted to female body extracts in Y-tube bioassays [30]. In addition, some studies with *D. maidis* in maize fields using sticky cards collected more males, which was attributed to males being caught in traps while searching for mates and attempting copulation [31,32]; even this cannot be confirmed only by sticky cards. In another study, the escape behavior of *D. maidis* was not affected by sex [33], indicating that the species exhibits variation in different aspects and warrants further investigation to better understand the searching behavior of males and females over both short and long distances. Our results may reflect short-distance attraction mediated by pheromone, since the distance from the odor source to the olfactometer was less than 50 cm.

The mating behavior of *D. maidis* has been previously studied, including sexual behavior and the role of vibrational and acoustic signals [9,10]. Notably, male behaviors such as wing fanning are more pronounced than those of females, which the authors believed might be associated with the release of medium-range chemical signals [9]. This supports our findings in the present study, showing that females are attracted to male odor even in the absence of visual or acoustic cues, demonstrating intraspecific chemical communication in this species. To the best of our knowledge, this is the first study to demonstrate the potential role of volatiles in sexual attraction within the superfamily Membracoidea, suggesting the presence of a pheromone in *D. maidis*.

The results obtained with non-acclimated females could suggest the existence of an alarm pheromone. This is supported by the observation that males preferred clean air over the odor of stressed female conspecifics, indicating avoidance of environments that may signal potential threats [34]. Leafhoppers may use chemical cues to mediate various behaviors, as it is common to observe aggregations of individuals on the same maize plant, similar to aphids, which rely on multiple chemical signals [35]. Likewise, aggregated treehoppers also release alarm pheromones, but only when their body wall is pierced, such as during a predator attack [36].

This study demonstrates that chemical communication plays a role in the reproductive behavior and attraction of *D. maidis*, alongside acoustic signaling. Given that insects must locate mates over long distances, it is reasonable to propose that many members of Auchenorrhyncha may also rely on chemical signals. These findings highlight the need for further investigation into chemical communication in this group.

Sexual pheromones are currently employed in various control strategies [37], including mass trapping with pheromone-baited traps, as is the case for *Tuta absoluta* (Meyrick) [38] mating disruption, which involves the release of synthetic pheromones in the field to reduce mate location and copulation rates [39], population monitoring [40–42] and attract-and-kill techniques [43]. Even in the absence of identified sex pheromones, chemical-based control strategies have been developed for other leafhopper species. For instance, a pushh–pull system using plant volatile compounds has been proposed for the tea green leafhopper *Empoasca flavescens* (Cicadellidae) [44,45]. Another system with *L. delicatula* showed effective trap attraction using pheromone lures based on body extracts to capture males and females during the oviposition period [46].

In this way, these findings are a step toward advancing knowledge of the chemical communication of *D. maidis* to understand their biology and ecology. Furthermore, it is also a possibility to develop new strategies to monitor or control this pest using chemical signals or extracts from living insects, specifically to capture females.

Supplementary Materials: The following supporting information can be downloaded at: https://www.mdpi.com/article/10.3390/insects16101021/s1, Figure S1: Male and female of *Dalbulus maidis*; Figure S2: Volatile sampling system for *Dalbulus maidis*.

Author Contributions: Conceptualization, M.S.S. and M.C.B.-M.; methodology, M.S.S., M.B., R.A.L., C.M.O., M.R.F. and M.C.B.-M.; validation, M.C.B.-M.; formal analysis, M.S.S. and M.C.B.-M.; investigation, M.S.S. and M.C.B.-M.; resources, M.B., R.A.L., C.M.O., M.C.B.-M. and M.R.F.; data curation, M.S.S. and M.C.B.-M.; writing—original draft preparation, M.S.S.; writing—review and editing, M.B., R.A.L., C.M.O., M.R.F. and M.C.B.-M.; supervision, M.R.F. and M.C.B.-M.; project administration, M.C.B.-M.; funding acquisition, M.B., R.A.L., C.M.O., M.C.B.-M. and M.R.F. All authors have read and agreed to the published version of the manuscript.

Funding: This work received financial support from the Coordination for the Improvement of Higher Education Personnel (CAPES) through a grant to MSS (process nr. 88887.939463/2024-00), the Brazilian Corporation of Agricultural Research (EMBRAPA, project N°. 20.18.04.004.00.00 and 20.19.00.182.00.00) and the Research Support Foundation of the Federal District (FAP-DF) with grant from 2021 to 2023 to MSS (project N°. 00193.00001056/2021-68), CNPQ project N°. 405307/2023-3.

Data Availability Statement: The raw data supporting the conclusions of this article will be made available by the authors, without undue reservation.

Acknowledgments: The authors would like to thank the National Council for Scientific and Technological Development (CNPq), the Coordination for the Improvement of Higher Education Personnel (CAPES), the Research Support Foundation of the Federal District (FAP-DF), and the Brazilian Agricultural Research Corporation (Embrapa) for financial and institutional support. We also thank Michely F.S. Aquino, Erika A.G. Jesus, Marisa L. Britto and Gabriel R.S. Macedo for helping with the insect colony.

Conflicts of Interest: The authors declare no conflicts of interest.

References

1. Whittaker, R.H.; Feeny, P.P. Allelochemics: Chemical interactions between species: Chemical agents are of major significance in the adaptation of species and organization of communities. *Science* **1971**, *171*, 757–770. [CrossRef] [PubMed]
2. Ali, M.F.; Morgan, E.D. Chemical communication in insect communities: A guide to insect pheromones with special emphasis on social insects. *Biol. Rev.* **1990**, *65*, 227–247. [CrossRef]
3. Yew, J.Y.; Chung, H. Insect pheromones: An overview of function, form, and discovery. *Prog. Lipid Res.* **2015**, *59*, 88–105. [CrossRef]
4. Nault, L.R. Evolution of insect pest maize and leafhopper, a case study. *Maydica* **1990**, *35*, 165–175.
5. Nault, L.R. Maize bushy stunt and corn stunt: A comparison of disease symptoms, pathogen host ranges and vectors. *Phytopathology* **1980**, *70*, 659–662. [CrossRef]
6. Villanova, E.S.; Ramos, A.; de Oliveira, M.C.S.; Esteves, M.B.; Gonçalves, M.C.; Lopes, J.R. First report of a mastrevirus (Geminiviridae) transmitted by the corn leafhopper. *Plant Dis.* **2022**, *106*, 1330–1333. [CrossRef]
7. Pérez-López, E.; Wist, T.; Rodríguez, Y.; Luna-Rodríguez, M.; Olivier, C.Y. Maize bushy stunt in native corn: Implications for Mexican "subsistence farmers". *Environ. Dev. Sustain.* **2018**, *20*, 1797–1805. [CrossRef]
8. Oliveira, C.M.; Frizzas, M.R. Eight decades of *Dalbulus maidis* (DeLong and Wolcott) (Hemiptera: Cicadellidae) in Brazil: What we know and what we need to know. *Neotrop. Entomol.* **2022**, *51*, 1–17. [CrossRef]
9. Heady, S.E.; Nault, L.R.; Shambaugh, G.F.; Fairchild, L. Acoustic and mating behavior of *Dalbulus* leafhoppers (Homoptera: Cicadellidae). *Ann. Entomol. Soc. Am.* **1986**, *79*, 727–736. [CrossRef]
10. Ramirez-Romero, R.; Perez-Ascencio, D.; Garibay-Benítez, D. Courtship behavior of the corn leafhopper *Dalbulus maidis* (DeLong & Wolcott) (Hemiptera: Cicadellidae). *J. Insect Behav.* **2014**, *27*, 804–815. [CrossRef]
11. Todd, J.L.; Phelan, P.L.; Nault, L.R. Interaction between visual and olfactory stimuli during host-finding by leafhopper, *Dalbulus maidis* (Homoptera: Cicadellidae). *J. Chem. Ecol.* **1990**, *16*, 2121–2133. [CrossRef]
12. Aráoz, M.C.; Jacobi, V.G.; Fernandez, P.C.; Albarracin, E.L.; Virla, E.G.; Hill, J.G.; Catalan, C.A.N. Volatiles mediate host-selection in the corn hoppers *Dalbulus maidis* (Hemiptera: Cicadellidae) and *Peregrinus maidis* (Hemiptera: Delphacidae). *Bull. Entomol. Res.* **2019**, *109*, 633–642. [CrossRef] [PubMed]
13. Chen, X.; Liang, A.P. Identification of a self-regulatory pheromone system that controls nymph aggregation behavior of rice spittlebug *Callitettix versicolor*. *Front. Zool.* **2015**, *12*, 1–12. [CrossRef] [PubMed]
14. Sevarika, M.; Rondoni, G.; Ganassi, S.; Pistillo, O.M.; Germinara, G.S.; De Cristofaro, A.; Romani, R.; Conti, E. Behavioural and electrophysiological responses of *Philaenus spumarius* to odours from conspecifics. *Sci. Rep.* **2022**, *12*, 8402. [CrossRef] [PubMed]
15. Witzgall, P.; Kirsch, P.; Cork, A. Sex pheromones and their impact on pest management. *J. Chem. Ecol.* **2010**, *36*, 80–100. [CrossRef]
16. Rizvi, S.A.H.; George, J.; Reddy, G.V.; Zeng, X.; Guerrero, A. Latest developments in insect sex pheromone research and its application in agricultural pest management. *Insects* **2021**, *12*, 484. [CrossRef]
17. Massola Júnior, N.S.; Bedendo, I.P.; Amorim, L.; Lopes, J.R.S. Effects of the inoculation time on corn with *Spiroplasma kunkelii* on yield components. *Fitopal. Bras.* **1999**, *24*, 4, 570–573.
18. Oliveira, C.M.; Sabato, E.O. *Doenças em milho: Insetos-vetores, molicutes e vírus*, 2nd ed.; Embrapa: Brasília, Brazil, 2017; pp. 153–184.
19. Sanches, M.S.; Michereff, M.F.F.; Borges, M.; Laumann, R.A.; Oliveira, C.M.; Frizzas, M.R.; Blassioli-Moraes, M.C. How much, how long and when: Density, duration and plant stage affect herbivore-induced plant volatiles in maize by the corn leafhopper. *J. Chem. Ecol.* **2025**, *51*, 53. [CrossRef]
20. NIST. National Institute of Standards and Technology. Mass Spectrometry Data Center. 2020. Available online: https://chemdata.nist.gov/ (accessed on 14 April 2025).
21. Leopold, E. Selective hydroboration of a 1, 3, 7-trene: Homogeraniol. *J. Org. Synth.* **1986**, *64*, 164–174.
22. R Core Team. R: A Language and Environment for Statistical Computing. 2016. Available online: https://www.r-project.org (accessed on 20 June 2025).
23. Faal, H.; Meier, L.R.; Canlas, I.J.; Murman, K.; Wallace, M.; Carrillo, D.; Cooperband, M.F. Volatiles from male honeydew excretions attract conspecific male spotted lanternflies, *Lycorma delicatula* (Hemiptera: Fulgoridae). *Front. Insect Sci.* **2022**, *2*, 982965. [CrossRef]
24. Moraes, M.C.; Borges, M.; Pareja, M.; Vieira, H.G.; Sereno, F.T.S.; Laumann, R.A. Food and humidity affect sex pheromone ratios in the stink bug, *Euschistus heros*. *Physiol. Entomol.* **2008**, *33*, 43–50. [CrossRef]
25. Claridge, M.F. Acoustic signals in the Homoptera: Behavior, taxonomy, and evolution. *Annu Rev Entomol* **1985**, *30*, 297–317. [CrossRef]
26. Virant-Doberlet, M.; Cokl, A. Vibrational communication in insects. *Neotrop. Entomol.* **2004**, *33*, 121–134. [CrossRef]
27. Strübing, H. *Sound Production: The Crucial Factor for Mate Finding in Planthoppers (Homoptera: Auchenorrhyncha)*; Studying Vibrational Communication; Springer: Cham, Switzerland, 2014; pp. 53–61.

28. Ranger, C.M.; Winter, R.E.; Backus, E.A.; Rottinghaus, G.E.; Ellersieck, M.R.; Johnson, D.W. Discrimination by the potato leafhopper (Hemiptera: Cicadellidae) of host volatiles from resistant and susceptible alfalfa, *Medicago sativa* L. *Environ. Entomol.* **2005**, *34*, 271–280. [CrossRef]
29. Rodrigues, I.; Benhadi-Marín, J.; Rodrigues, N.; Baptista, P.; Pereira, J.A. Olfactory responses to volatile organic compounds and movement parameters of *Philaenus spumarius* and *Cicadella viridis*. *J. Appl. Entomol.* **2022**, *146*, 486–497. [CrossRef]
30. Faal, H.; Cooperband, M.F.; Canlas, I.; Carrillo, D. Evidence of pheromone use in a fulgorid, spotted lanternfly. *Forests* **2022**, *13*, 1639. [CrossRef]
31. Oliveira, C.M.; Lopes, J.R.; Nault, L.R. Survival Strategies of *Dalbulus maidis* During Maize Off-Season in Brazil. *Entomol. Exp. Appl. 2013 147*, 141–153. [CrossRef]
32. Canale, M.C.; Lopes, J.R.S.; Nesi, C.N.; Prado, S.d.S. Role of Dalbulus maidis (Hemiptera: Cicadellidae) gender on maizebushy stunt phytoplasma transmission. *Phytopathogenic Mollicutes* **2018**, *8*, 32–39. [CrossRef]
33. Oliveira, C.M.; Querino, R.B.; Jesus, E.A.G.; Sanches, M.S.; Aquino, M.F.S.; Souza, H.R.; Brito, M.L.; Malaquias, J.V.; Conceição, G.V.; Frizzas, M.R. Escape Behaviour of *Dalbulus maidis* (DeLong and Wolcott): Subsidies for Insecticide Application and Field Sampling. *J. Appl. Entomol.* 2025. [CrossRef]
34. Verheggen, F.J.; Haubruge, E.; Mescher, M.C. Alarm pheromones—Chemical signaling in response to danger. *Vitam. Horm.* **2010**, *83*, 215–239.
35. Boullis, A.; Verheggen, F.J. Chemical Ecology of Aphids (Hemiptera: Aphididae); Biology and Ecology of Aphids. In *Biology and Ecology of Aphids*; CRC Press: Boca Raton, FL, USA, 2016; p. 171.
36. Nault, L.R.; Wood, T.K.; Goff, A.M. Treehopper (Membracidae) alarm pheromones. *Nature* **1974**, *249*, 387–388. [CrossRef] [PubMed]
37. Blassioli-Moraes, M.C.; Laumann, R.A.; Michereff, M.F.; Borges, M. Semiochemicals for Integrated Pest Management. In *Sustainable Agrochemistry: A Compendium of Technologies*; Peres, F., Lichtenstein, G., Eds.; Springer: Cham, Switzerland, 2019; pp. 85–112.
38. Cocco, A.; Deliperi, S.; Delrio, G. Potential of mass trapping for *Tuta absoluta* management in greenhouse tomato crops using light and pheromone traps. *IOBC-WPRS Bull.* **2012**, *80*, 319–324.
39. Hasan, M.M.; Mahroof, R.M.; Aikins, M.J.; Athanassiou, C.G.; Phillips, T.W. Pheromone-based auto-confusion for mating disruption of *Plodia interpunctella* (Lepidoptera: Pyralidae) in structures with raw and processed grain products. *J. Stored. Prod. Res.* **2023**, *104*, 102201. [CrossRef]
40. Kim, Y.; Jung, S.; Kim, Y.; Lee, Y. Real-time monitoring of oriental fruit moth, *Grapholita molesta*, populations using a remote sensing pheromone trap in apple orchards. *J. Asia Pac. Entomol.* **2011**, *14*, 259–262. [CrossRef]
41. Laumann, R.A.; Moraes, M.C.B.; Khrimian, A.; Borges, M. Field capture of *Thyanta perditor* with pheromone-baited traps. *Pesq. Agropec. Bras.* **2011**, *46*, 113–119. [CrossRef]
42. Silva, V.P.D.; Pereira, M.J.B.; Vivan, L.M.; Blassioli-Moraes, M.C.; Laumann, R.A.; Borges, M. Monitoring of the brown stink bug *Euschistus heros* (Hemiptera: Pentatomidae) with sex pheromone in soybean fields. *Pesq. Agropec. Bras.* **2014**, *49*, 844–852. [CrossRef]
43. Krupke, C.H.; Roitberg, B.D.; Judd, G.J.R. Field and laboratory responses of male codling moth (Lepidoptera: Tortricidae) to a pheromone-based attract-and-kill strategy. *Environ. Entomol.* **2002**, *31*, 189–197. [CrossRef]
44. Zhang, Z.; Luo, Z.; Gao, Y.; Bian, L.; Sun, X.; Chen, Z. Volatiles from non-host aromatic plants repel tea green leafhopper *Empoasca vitis*. *Entomol. Exp. Appl.* **2014**, *153*, 156–169. [CrossRef]
45. Niu, Y.; Han, S.; Wu, Z.; Pan, C.; Wang, M.; Tang, Y.; Zhang, Q.H.; Tan, G.; Han, B. A push–pull strategy for controlling the tea green leafhopper (*Empoasca flavescens* F.) using semiochemicals from *Tagetes erecta* and *Flemingia macrophylla*. *Pest Manag. Sci.* **2022**, *78*, 2161–2172. [CrossRef]
46. Cooperband, M.F.; Murman, K.M. Spotted lanternflies respond to natural pheromone lures for mate-finding and oviposition. *Insects* **2024**, *15*, 447. [CrossRef]

MDPI AG
Grosspeteranlage 5
4052 Basel
Switzerland
Tel.: +41 61 683 77 34

Insects Editorial Office
E-mail: insects@mdpi.com
www.mdpi.com/journal/insects

www.ingramcontent.com/pod-product-compliance
Lightning Source LLC
Chambersburg PA
CBHW062052220926
51850CB00010B/2744